Settlement Ecology of the Early and Middle Neolithic Körös and Linear Pottery Cultures in Hungary

Krisztina Kosse

I0894685

BAR International Series 64
1979

B.A.R.

B.A.R., 122 Banbury Road, Oxford OX2 7BP, England

GENERAL EDITORS

A. R. Hands, B.Sc., M.A., D.Phil.
D. R. Walker, M.A.

B. A. R. -S64, 1979: "Settlement Ecology of the Early and Middle Neolithic Körös and Linear Pottery Cultures in Hungary".

© Krisztina Kosse, 1979.

The author's moral rights under the 1988 UK Copyright, Designs and Patents Act are hereby expressly asserted.

All rights reserved. No part of this work may be copied, reproduced, stored, sold, distributed, scanned, saved in any form of digital format or transmitted in any form digitally, without the written permission of the Publisher.

ISBN 9780860540663 paperback
ISBN 9781407349589 e-book
DOI https://doi.org/10.30861/9780860540663
A catalogue record for this book is available from the British Library

This book is available at www.barpublishing.com

CONTENTS

Page

LIST OF FIGURES

LIST OF TABLES

MAPS

Please note that additional material is available to download from www.barpublishing.com/additional-downloads.html. The original foldout has been reduced in size to match the A4 format of this book, the image is therefore not as clear as the original foldout. Please refer to the original foldout via the download for the original content.

ACKNOWLEDGEMENTS

This study was originally presented as a thesis for the degree of Ph.D. to the University of London, Institute of Archaeology. With minor modifications the text and illustrations are essentially those of the thesis. The completion of the thesis would have been impossible without the help of many people, both in England and Hungary. To them I would like to express my gratitude.

I would like to thank Dr. Nándor Kalicz and Dr. János Makkay, Institute of Archaeology, Budapest, for devoting so much of their time to help me locate the sites, for showing me sites they have excavated or surveyed, and for the long discussions we had on the archaeology and ecology of the Körös and Linear Pottery cultures. I am especially grateful to Dr. János Makkay for reading the completed thesis and for the corrections he has suggested.

I also received help from archaeologists of many regional museums in Hungary: Dr. Gábor Bándi (Museum of Szombathely), István Ecsedy (Museum of Pécs), György Goldman (Museum of Békéscsaba), Katalin Hegedüs (Museum of Szentes), Ferenc Horváth (Museum of Szeged), László Horváth (Museum of Nagykanizsa), Dr. József Korek (National Museum of Hungary) and Dr. Otto Trogmayer (Museum of Szeged).

For information from other disciplines, I would like to thank Dr. Zoltán Borsy (Dept. of Geography, University of Debrecen), Dr. Miklós Füzes (Museum of Keszthely), Dr. Borbála Hartyányi and Dr. Gyula Nováki (Agricultural Museum, Budapest), Dr. Magda Járai-Komlódi (Dept. of Plant Taxonomy and Geography, Eötvös Loránd University, Budapest), Dr. Márton Pécsi (Hungarian Academy of Sciences), Dr. Sándor Somogyi and Dr. Sándor Marosi (Institute of Geography, Budapest), Dr. János Stieber (Institute of Applied Botany, Budapest), Dr. Lajos Takács (Ethnographical Museum, Budapest), Dr. György Várallyay (Institute of Agrochemistry and Soil Science, Budapest) and Dr. Bálint Zólyomi (Botanical Research Institute, Vácrátot).

In England, I am especially grateful to Prof. J. D. Evans, University of London, Institute of Archaeology, for his help, advice and encouragement during the preparation of the thesis. I would also like to thank Dr. J. G. Nandris, University of London, Institute of Archaeology, for help, advice and discussions concerning the work.

I am also grateful to Prof. M. Stone, University College, London, for advice and help with statistical techniques and interpretation.

I am indebted to the Neale and Eveleth Gift Foundation for financial support for four years, and to the British Council, whose Cultural Exchange Programme enabled me to spend two months in Hungary.

Finally, I would like to thank my husband, Alan, who, when he could least afford it, took time off to travel to Hungary to collect some additional information for my thesis. I am also grateful for the many references he has given me, for the long and useful discussions we had, and for encouragement and support during the work.

INTRODUCTION

Agriculture based on wheat and barley cultivation and sheep/goat husbandry was according to the evidence of both archaeology and the natural sciences introduced to Europe from the Near East. There is increasing reason to believe that manipulation of local resources resulting in some measure of crop and herd control was a widespread phenomenon in late Postglacial Europe (Clarke, 1976). Nonetheless, cereal agriculture with its high yielding and easily storable annuals together with sheep and goat husbandry proved more productive and replaced or absorbed the old practices into the new economy. Radiocarbon dates indicate that by 6000 bc, cereal agriculture had become well established in Greece, where we also have the earliest evidence for the domestication of cattle (Boessneck, 1962; Higgs, 1962). By about 5000 bc, agricultural settlements had existed in Europe as far north as the Great Hungarian Plain, and by 4000 bc agriculture had reached Holland (Clark, 1965). By 3000 bc, cereal cultivation and sheep/goat husbandry augmented with some new local domesticates had become the dominant food producing system in most areas of temperate Europe.

Although Near Eastern agriculture was potentially more productive than hunting and gathering or other forms of indigenous resource manipulation, it was also initially ill-adapted to the environmental conditions of temperate Europe. The Near Eastern nuclear area, including the habitats of wild cereals and the area of early agricultural settlements, lies in the subtropical zone. The climate is characterised by a winter precipitation maximum, hot and dry summers and a high percentage of the possible sunshine for the year (Trewartha, 1954). The growing season of the cereals and all early Near Eastern crops is a winter one (Butzer, 1971; Oates and Oates, 1976). The amount of rainfall varies according to altitude, but in areas more suitable for agriculture, it is low and often unpredictable. Water is a critical resource for agriculture. Van Zeist (1969), on the basis of pollen evidence, suggests that summers were probably drier than today in the Near East during the early spread of agriculture, although he sees little difference in the amount of annual precipitation. The highest concentration of early farming settlements in both the lowland steppe and the oak-pistachio belt is found in areas of high water table, often at swamp margins (Flannery, 1971). The first farming settlements outside the Near Eastern nuclear area are found in Greece, which is still in the winter rainfall belt. The introduction of agriculture further north necessitated the adaptation of the original western Asiatic domesticates to the environmental conditions of a new climatic zone.

The first introduction of agriculture to temperate Europe is associated with the Karanovo-Starčevo-Criş-Körös complex and the Linear Pottery culture. The distribution area of the Karanovo complex includes much of Bulgaria, eastern and central Yugoslavia, Rumania and southeastern Hungary. Although the maximum precipitation is in summer and winters are cold, southeast Europe due to its low latitude enjoys a climate intermediate between the

Mediterranean and the Temperate (Childe, 1957). Much of the Balkan peninsula is still subhumid; the climate is characterised by abundant sunshine, a long growing season and a large average rainfall deficit in summer (Trewartha, 1954; Thran and Broekhuizen, 1965). Despite the presence of some new features such as the gabled roof, the use of wood in house construction and the detached house, the material culture of the first farming communities of southeast Europe was similar to that of western Asia. Southeast Europe was still essentially a western province of the Near Eastern culture area (Piggott, 1965; Tringham, 1971). The adaptation of agriculture to the wetter forested environments of midlatitude temperate Europe was the achievement of the Linear Pottery culture (Clark and Piggott, 1965; Nandris, 1972).

The available evidence places the earliest appearance of the Linear Pottery culture in the area of modern Hungary, the northernmost extension of the southeast European culture area. Quitta (1964) suggested that the origins of the Bandkeramik should be sought in the fertile valleys of hilly Transdanubia, and Waterbolk (1971, 347) considered the Hungarian Plain as the dominant centre 'which had incorporated all the major achievements of the nuclear area in the Near East, and which had added to these an adaptation to the European deciduous forest'. The discovery of the so called Szatmár group in the Great Hungarian Plain which provides a material link between the Körös culture and the Alföld Linear Pottery culture, and that of the Medina group which constitutes a similar link between the Starčevo and Transdanubian Linear Pottery cultures would indicate that both these areas were important in the genesis of the Linear Pottery culture (Kalicz and Makkay, 1972a; 1972b). While it is not expected that Hungary would reflect all the ecological changes that accompanied the introduction of agriculture to Europe, its position as a transitional zone suggests that a detailed study of the Körös and Hungarian Linear Pottery adaptations would throw much light on how cereal farming and animal husbandry were adapted to temperate Europe.

In this study, an attempt is made to identify the changes that occurred in man-land relationships in Hungary during the spread of agriculture through an examination of the settlement ecologies of the Körös, Alföld- and Transdanubian Linear Pottery cultures. Archaeological evidence by its nature is of only limited usefulness in discovering prehistoric man-land relationships; but, as some recent ecological-archaeological studies conducted in different parts of the world have demonstrated, when information from the environmental sciences is used in conjunction with the archaeological evidence, a tentative reconstruction of past land use patterns should be possible. In the following study, the environmental information is obtained from two main sources: 1) plants and animal remains from the sites; and 2) the analysis of settlement distributions in relation to climatic areas, vegetation, hydrography, geology and soils.

Previous Investigations

The geographical approach in archaeology has a long tradition in Europe. Distribution maps showing the provenance of antiquities began to be published around the middle of the last century, and as early as 1874 the International Congress at Stockholm adopted a standard set of international symbols to be used in archaeological mapping (Daniel, 1950). It was not, however, until the emergence of the anthropogeographical and, somewhat later, Kuturkreis schools in Germany and Austria that the careful mapping of artifact and trait distributions became an important feature of archaeological methodology. Settlement geography, closely related to the anthropogeographical school, emerged as a subdiscipline of cultural geography around the beginning of this century (Sauer, 1962). Its two leading exponents, Gradmann and Schlüter, advocated not only the study of settlement distributions in relation to the earth surface, but also the reconstruction of former conditions of settlement and land use.

In his 'Steppenheide' theory, Gradmann (1906; 1933) put forward the general thesis that neolithic and bronze age settlement in Central Europe was originally confined to open country of steppe- or heath-like character. The basis for the theory was partly the observation of a close correspondence between prehistoric settlement and the recent distribution of steppe- and heath-like vegetation, and partly the assumption that early farming man was incapable of clearing the forest.

Since then, pollen studies have effectively disproved Gradmann's theory. They have shown, that contrary to his opinion, the loess lands at the time of early neolithic settlement were mostly covered by forest vegetation (Garnett, 1945; Firbas, 1949-52). Experiments with neolithic stone axes have proved that neolithic man had the technology which would have enabled him to fell trees and clear forests (Iversen, 1956). The importance of Gradmann's work lies not in the validity of his hypothesis, but in the impact it had on prehistoric archaeology. By stimulating interest in the environmental setting of archaeological settlements, he opened up a new field of enquiry. Settlement geography, as used by Gradmann and his school, also provided a method of analysis.

Since the war, there has been a number of studies aimed at elucidating the physical environment of the Linear Pottery and the Karanovo-Körös settlements. The approach has generally been to analyse the distribution of settlements in relation to one major environmental variable.

In 1948, Schwarz used Gradmann's ecological-distributional method to disprove the 'Steppenheide' theory. Schwarz examined the distribution of the Linear Pottery settlements against the distribution of chernozem soils in central Germany. In answer to the question set out in the title of his work, 'Lagen die Siedlungen der linearbandkeramischen Kulturs Mitteldeutschlands in waldfreien oder bewaldeten Lanschaften?', Schwarz has shown that the Bandkeramik settlers instead of seeking the chernozem areas with their open grassy vegetation, seemed to have avoided them. Wilhelmy (1950) comes to the same conclusion in regards to the distribution of Bulgarian prehistoric settlements. Wilhelmy's article, whose main concern is the age of the central and eastern European chernozem soils, is a skillful treatment of archaeological evidence in an environmental study.

Another field of interest has been the evident correlation of Bandkeramik settlements with the loess areas of Europe. Narr (1956) used the distributional method to discover why certain loess deposits in the northern Rhineland attracted Bandkeramik settlements, while others remained empty. He demonstrated that where the Danubians settled the loess cover was less thick and the brown earth had a higher degree of basal saturation. Modderman (1958-59) posed the same question in relation to the Dutch Bandkeramik sites. His findings are slightly different from those of Narr's. The sites along the Maas are found on the second terrace of the river, where the loess cover is especially thick. Modderman attributes this to differences in drainage conditions. In Holland and Belgium, those loess areas were selected for settlement where the ground water table lay sufficiently deep under a thick loess cover to ensure good natural drainage and where consequently the soils were also easy to work. Modderman describes the ideal area of Bandkeramik settlement in western Europe as 'ein flacher, entwässerter Lössboden in unmittelbarer Nahe des Wassers'. According to Quitta (1970), most Bandkeramik settlements in the Leipzig region are also located on well drained soils. Here, however, the settlers preferred the lower terraces, where good drainage is provided by morainic sand or gravel under a relatively thin loess cover. Quitta, similarly to Schwarz and Wilhelmy, notes the absence of Bandkeramik settlements from the drier loess plateaus; the sites are found along the edge of the chernozem zone, mostly on base rich parabraunerden.

Nandris (1970: 1972) has emphasised the importance of good drainage in the selection of the Körös sites. The settlements along the Körös and Tisza rivers are located in those areas where the ground water table lies deep under the surface. In southeast Europe, however, as Sherratt (1972; 1973) has shown, soil moisture was, on the whole, still a critical factor. Early neolithic sites in Bulgaria and Yugoslavia, as in Greece and Turkey, are located in fertile alluvial basins or river valleys in or adjacent to areas of high ground water table.

In recent years, there has been a growing number of studies aimed at a more total reconstruction of past human ecologies. Most of these have been undertaken in areas of primary domestication, in the Near East and Mesoamerica. Nevertheless, two studies have dealt directly with the Linear Pottery culture. The first one, by Sielmann, has been concerned with Linear Pottery adaptation in Germany, and the second, by Kruk, with Linear Pottery adaptation in Poland. Both studies are based on an analysis of settlement distributions.

Sielmann (1971; 1972) examined the distribution of Linear Pottery settlements in relation to parent material and climate. He concluded that from the beginning of the Bandkeramik development there had been two settlement zones. Ecology area A, comprising central Germany, the Lower Main region, the northern Rhenish plain, the area east of Mannheim and the central Rhine region, is characterised by relatively low rainfall, high temperature and consequently by higher aridity. Ecology area B, which comprises the Bavarian loess region, central Neckerland and the Heidelberg area, is characterised by either high rainfall and high temperatures or by average rainfall and lower temperatures, and consequently by lower aridity. Sielmann suggested that these two ecology areas were the results oftwo waves of colonisation: Ecology area A was colonised by settlers coming from a drier part of the Starčevo-Körös area, and Ecology area B by settlers from a wetter region of the Starčevo culture. The two settlement zones differed somewhat in their economies and also in their material cultures. In the wetter areas, cattle were far more frequent among the domesticated animals and Sielmann assumes that stockbreeding would have been more important than cereal agriculture. Sielmann's study is interesting not only because he demonstrates his findings by quantitative data, but also because he uses climate as an independent variable.

On the other hand, there is little attempt at an overall ecological analysis; the relationship of settlements to vegetation, hydrography and soils is mentioned, but not elaborated in detail.

Kruk's (1971; 1972-73) study is an analysis of the interdependence of neolithic settlement, soil types and vegetation in the loess region of the Nida basin in Poland. He shows that the areas occupied by the Linear Pottery culture were covered by brown-loessic soils and carried a dense forest cover. In this, his findings are similar to Schwarz's and Wilhelmy's. Settlements were mostly situated in lower lying areas, near rivers: the loess uplands were only sporadically utilised. Kruk emphasises man's influence on the landscape. In the genesis of chernozem soils, in contrast to Wilhelmy, he considers anthropogenic influence a major factor. He attributes the development of these soils in Poland to the slash and burn agriculture and large scale pastoral activities of the TRB and Corded Ware cultures. He thinks, even so, that the Danubian economy had not seriously disturbed the ecological balance of the area. Kruk's study is a careful evaluation of the mutual effects that existed between man and the environment. However, while the relationships discovered in Poland might, as claimed by Kruk, apply to most of the Middle European loess zone, the same assumption cannot automatically be made for the loess zones of southeast Europe. In the following study, it will be suggested that in Hungary, at least, it is unlikely that anthropogenic influence was the sole factor in the creation of chernozem soils.

Geographical-Ecological Approach

The study of man-land relationships is, by its nature, an essentially ecological problem. Ecology may be defined as the interrelationships between living organisms and their biotic and physical environments (Ford, 1972). The biotic environment comprises the whole complex of living organisms—plants, animals and men, and the physical environment—the soil, lithology and climate (Clark, 1952; Eyre, 1963). When dealing with human communities, however, man's cultural environment need also be considered. The relevant or effective environment will in fact depend upon the culture. As Steward (1955, 40) points out: 'The simpler cultures are more directly conditioned by the environment than advanced ones. In general, climate, topography, soils, hydrography, vegetation cover, and fauna are crucial, but some features may be more important than others. The spacing of water holes in the desert may be vital to a nomadic seed-gathering people, the habits of game will affect the way hunting is done, and the kinds and seasons of fish runs will determine the habits of riverine and coastal tribes'. Soil fertility will be of little direct consequence for hunter-fishers, but will be a major consideration for agriculturalists. Changes in technology or some other aspect of the culture can alter the effective environment without any change in the physical environment. The reconstruction of past man-land relationship requires both an analysis, the discovery of the relevant environmental features, and a synthesis, the integration of 'natural' and 'cultural' data.

In the first part of the following study, an attempt is made to delineate the natural environment which was deemed optimal by the peoples of the Körös and Linear Pottery cultures for the functioning of their economies. To achieve this end, the distributions of the Körös and Linear Pottery settlements are examined in relation to basic elements of the environment. The method is the geographical method of the previous studies. An alternative approach, using the site as the focal point of investigations has been advanced recently by Vita-Finzi and Higgs (1970; Higgs and Vita-Finzi, 1972). But while the method of site-catchment analysis is useful for the study of man's changing relationship with his immediate local environment, it is less suited for the study of larger regional patterns, with which this thesis is concerned.

A proposition which is basic to both approaches is that sites were located so as to minimise the effort needed for acquiring critical resources. The 'law of minimum effort' first developed by Lösch (cf. 1954) and translated into the concept of 'least movement' has become an important explanatory model in locational geography. As Haggett (1965, 33) suggests 'the actual paths will tend to fluctuate about the optimum in a random way' but they will tend to give a distribution around the least effort path. In behavioural terms, it seems a reasonable assumption, and is also borne out by empirical data, that people rarely choose to do more work than is necessary (Chisholm, 1968). While some of the rationale of prehistoric peoples may escape us, it is unlikely that practical considerations would have been totally irrelevant.

Studies on 'natural regions' suggest that although an undoubtedly close correlation exists between elements of the environment, the boundaries of soil, climate and vegetation do not always coincide (Griff, 1967). Because of this, the relationships of the settlements to climate, vegetation, hydrography, geology and soils will be examined separately.

The second part of the study is an attempt at integration: the environmental evidence, its limitations and potentialities, are correlated with the available archaeological evidence, especially floral and faunal data, in a reconstruction of the Körös and Linear Pottery economies. In the interpretation, considerable use is made of ethnohistorical parallels from the area. While the indiscriminate use of ethnohistoric evidence is justly considered misleading, its careful use can provide important insights into aspects of prehistoric life for which no other evidence exists (Clark, 1952). In Hungary, where ethnohistorical research has had a long tradition and where industrialisation has been relatively late, there is a large body of carefully collected data providing a valuable source of information about past land use practices.

METHODS

This section describes the procedures employed in mapping the sites, and the statistical tests used to check the associations between settlements and environmental variables. The sites included in the analysis, as the archaeological evidence would suggest, are all habitation sites (Trogmayer, 1968; Kalicz and Makkay, 1977). Because of the lack of adequate data, in the initial analysis no distinction is made between the settlements according to size or the length of the occupation. These questions will, however, be discussed in the second part of the study.

Location of Sites

Sites were first located on sheets of the 1:200 000 Generalkarte von Mitteleuropa series. Because of the obvious need for precision, only sites that could be located at this scale were included in the subsequent analysis. This meant 131 Körös, 385 Alföld Linear Pottery, and 212 Transdanubian Linear Pottery sites. As regards the rest of the sites, in some cases the location was entirely unknown; in others, the information was inadequate for accurate location. The site list given in Appendix A includes all the sites; Map 1 shows the sites included in the analysis. The numbers on the map refer to the numbers given to the sites in the Appendix.

The analysis was carried out on 1:500 000 maps. It was found that at this scale the maps provided enough detail for the purposes of the investigation, and that the sites could be plotted without too much overlapping. All the maps, with the exception of that on hydrography, were obtained from the National Atlas of Hungary (1967). The hydrography map is a composite of several older maps prepared by the Hydrography Institute of the Ministry of Agriculture, Budapest (1938). The maps included here are approximately 1:4 of the 1:500 000 base maps.

Statistical Methods

Before describing the tests used in the analysis, it should be emphasised that the sample is a non-random one: the sites selected were those which could be accurately located. In addition, site discovery itself is a non-random process, as certain areas will be less well represented than others (Clarke, 1968). Because of these inherent limitations in the sample, the results should be regarded more in the nature of tentative suggestions rather than definitive statements. In the thesis, statistical methods were used to express aspects of the relationship between settlements and the environment which would not have been explicit from the inspection of the data.

Two tests were involved in checking the associations. The first one, chi-square, was employed to determine whether the observed distributions differed significantly from distributions which might have been expected by chance alone. The second one, 'percentage point difference', was used to assess the relative importance of the different variables.

MAP 1. Distribution of the Kőrős and Linear Pottery Settlements in Hungary

Key

Elevation above sea level

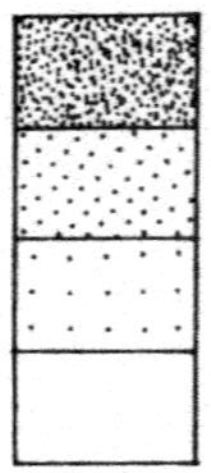

■ Settlements of the Kőrős culture

▲ Settlements of the Linear Pottery culture in Eastern Hungary

● Settlements of the Transdanubian Linear Pottery culture

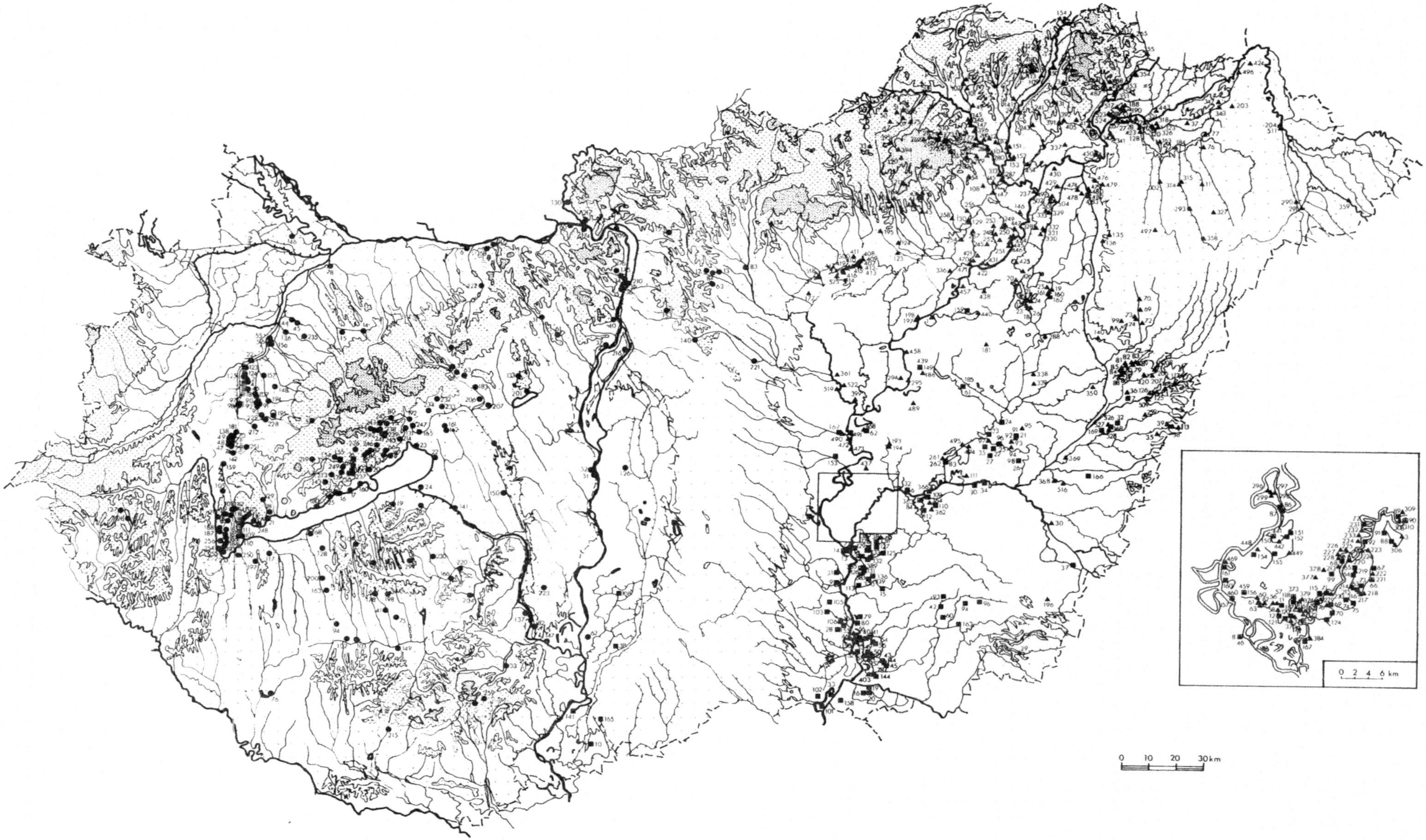

0 2 4 6 km
0 10 20 30 km

This latter method is a simple statistic, in which the proportion of the variable in the universe is compared with the proportion observed in the sample (Plog and Hill, 1971). If the variable had no effect on site distribution, the percentage in the sample would be similar to the percentage of the variable in the universe. The difference between the two proportions measures the effect of the variable.

The percentages of the variables in the universe (in this case Hungary) were calculated with the help of a grid system, although other methods can also be used. A 5 mm square grid was superimposed on each environmental map at a scale of 1:1 000 000, and the variable in the centre of the square was recorded. The percentage of each variable in Hungary was calculated on the basis of 3681 readings (the number of squares in the grid). To check whether the mesh was fine enough, the same calculation was performed using a grid of 1 mm squares. The difference between the results obtained by the two grids was less than a fraction of one percent.

To discover the effect of climate, soil, etc. on settlement distribution, the importance of each class of variables was calculated separately. As, however, the joint effect of several variables may be expected to be greater than their independent actions (Plog, 1971), the combined effect of the environmental elements was also measured. To achieve this end, the variables used in the separate analyses were reduced to 14 (Appendix B). The number of possible combinations for the five classes (geology, climate, soils, vegetation, and distance from rivers) of 14 variables is 120. The percentage of each combination occurring in Hungary was calculated as before, using the grid method, except that instead of one variable 5 different types of variables were recorded for each square. The percentages for each combination in the universe were then compared with the percentages obtained for the sites. The method is a simple one, although it involves a large number of calculations. The use of a computer could reduce the time spent on the analysis, and at the same time by allowing an increase in the number of the variables, it could also extend the explanatory potential of the method.

The idea of using regression analysis to measure the simultaneous effect of the variables was considered, especially as the method has already been used in a similar archaeological study (Green, 1973). However, some of the variables would have been difficult to quantify, and, furthermore, the statistical interpretation of the results would have raised a number of problems (Prof. Stone, pers. comm.). The use of a simpler, though less powerful method, seemed more suitable in view of the nature of the data.

CHRONOLOGY

The first introduction of farming to Hungary is associated with the Körös culture, the northernmost extension of the Karanovo-Criş complex. In 1944, Kutzián already recognised that the Körös culture was genetically linked with the Starčevo culture of Serbia and also suggested that both cultures appeared to be intrusive in their respective areas. Research since then has confirmed her view; it has also extended the relationship to include the Karanovo and Kremikovci cultures of Bulgaria and the Criş culture of Romania. The material attributes of the complex have been described by Tringham (1971).

The distribution area of the Körös culture, marked in the north by the Tisza river, the towns of Kunhegyes, Karcag and Berettyóujfalu, includes the southern part of the Alföld. The largest concentration of sites is in modern Hungary, but they are also found in northern Yugoslavia and parts of north-western Romania. While the areal extent of the Körös culture is relatively small, it does include a large number of sites. As Nandris (1970) has pointed out,about a third of the known sites in the Karanovo complex, are in the Körös area. At present, the number of known Körös sites in Hungary is around 169 (Appendix A), but the indications are that this number will be substantially increased when the intensive archaeological survey of Hungary, which is currently being carried out by the Institute of Archaeology of Budapest, is finally accomplished (Makkay, pers. comm.). For practical reasons, the emphasis in this thesis will be on the Hungarian sites, although references will be made to sites south of the border.

The first detailed chronology of the Starčevo culture was devised by Milojčić (1949; 1950). Basing his scheme on the material of the eponymous site, he divided the Starčevo culture into four developmental phases and suggested that the Körös culture was parallel with the last phase. Subsequent research has considerably revised Milojčić's chronology. Hungarian archaeologists consider the Körös culture to have been contemporaneous with most of the Starčevo development (Trogmayer, 1968; Makkay, 1969a). Radiocarbon dates would tend to confirm this view: the earliest Körös site is dated to 5140 bc and the latest one to 4420 bc (Kalicz and Makkay, 1977). This would indicate some 700 years of local evolution, and therefore a longer chronology than that envisaged by Milojčić. The dates are comparable to those obtained for the Karanovo and Starčevo cultures (Tringham, 1968) and seem to be acceptable in terms of the material evidence to most archaeologists. Some of the early misunderstanding over chronology has arisen from the belief that the Körös culture entirely lacked painted pottery. The erroneous nature of this conception has been discussed in detail by Makkay and Trogmayer (1966). It is now known that the painted ware of the Körös culture includes white, red and black painted decoration. Nonetheless, in the absence of tell settlements in Hungary, the internal chronology of the Körös culture is still somewhat problematic.

Recently, Trogmayer (1968a) has attempted to establish a temporal sequence for the Körös sites on the basis of the proportion of barbotine decoration. The earliest phase, represented by Gyálarét and Röszke-Ludvár, is characterised by an absence of barbotine decoration. In the following phases, Gyálarét pit 8, Maroslele-Pana pit 4, Deszk pits 13, 8, and 7, the proportion of barbotine ware gradually increases to reach a maximum in the last phase: Pana pit 3 and Deszk pit 15. The youngest phase is also distinguished by the appearance of certain Vinča traits such as biconical bowls and fluted decoration. According to Trogmayer, the same succession can be observed at Vršnik, where chronology is supported by stratigraphical evidence. The high percentage of barbotine ware which characterises all the finds at Starčevo, would suggest that the site on which Milojčić based his chronology was occupied only during the youngest phase of the culture's development. The validity of Trogmayer's chronology seems to be borne out not only by the parallel evolution shown at Vršnik, but also to some extent by radiocarbon dates. However, his scheme only includes a very few sites: the chronological position of the majority of the settlements has still to be determined.

In contrast to the Körös culture, the internal chronology of the Alföld Linear Pottery culture (ALP) is relatively well established. The earliest phase, referred to as the Szatmár group, is essentially a phase of transition from the Körös to the Linear Pottery culture. In Szatmár I, Körös elements are still preponderant, but in Szatmár II, their number decreases. The sites are found outside the Körös area, mostly at the northern and eastern edge of the Alföld. Outside Hungary, Michalovce-Hrádok in southeastern Slovakia and Ciumeşti-Berea IX in Romania belong to this phase. Lichardus (1972) refers to this earliest phase as the Protobandkeramik Horizon. An interesting recently excavated site in Hungary is Méhtelek. Its material is somewhat atypical, as it shows connections with the earliest neolithic of Transylvania rather than the Alföld. The date for Méhtelek, 4675 bc (Kalicz and Makkay, 1974), also gives an approximate date for the Szatmár group. Kalicz and Makkay consider the Szatmár group to have been parallel with the last phase of the Körös culture. The incised sherds found at Gyálarét and Röszke they regard as atypical and not representative of the Linear Pottery culture.

The second phase of the ALP is characterised by developed Linear Pottery forms and decorations: Körös traits are entirely absent. The sites are still concentrated in the northern and eastern parts, although some are already in the Körös distribution area. The last phase is distinguished by the development of local groups. The earliest one is Tiszadob (Lichardus's Gemör group), which is the forerunner of the Bükk culture. The Tiszadob group is divided into two, and the Bükk culture into three developmental phases. Bükk II or Classical Bükk marks the highest stage of the Linear Pottery culture in the Alföld, with extensive connections outside Hungary. The other groups are: Late AVK, Esztár, Szilmeg and Szakálhát. The Szakálhát group, which occupies the old area of the Körös culture, shows strong links with Vinča in Yugoslavia. The groups were largely contemporaneous as suggested by their sherds on each other's sites. Good cross dating also exists with the Transdanubian Notenkopf and Zseliz phases. In terms of the Vinča development, the ALP may be equated with Vinča A and B1-B2. Radiocarbon dates are still very few: they range from 4670 to 4130 bc. The material culture of the ALP has been described in detail by Kalicz and Makkay (1977).

Please note that a full-size version of this map is available to download from www.barpublishing.com/additional-downloads.html
The original foldout has been reduced in size to match the A4 format of this book, the image is therefore not as clear as the original foldout. Please refer to the original foldout via the download for the original content.
Please also be aware that the image might be cut off or not complete in the printed book.

The internal chronology of the Transdanubian Linear Pottery culture (TLP) is somewhat less secure. The earliest phase, referred to as the Medina group, seems to be an equivalent of the Szatmár group (Kalicz and Makkay, 1972; 1972b). It has a very strong Starčevo connections and may in fact be regarded as a local variant of the Starčevo-Körös group. The next phase, Very Early TLP, is represented so far by only a very few sites: Bicske, Révfülöp, Vonyarc-Vashegy, etc. (Kalicz and Makkay, pers. comm.). Outside Hungary it is found at Bíňa in southwest Slovakia. On the basis of biconical bowls and fluted decoration at Bíňa, Pavúk (1973) equates this phase with the Protovinča horizon of the Late Starčevo-Körös culture. The fluted decoration on linear pottery is not found outside the Carpathian basin.

The next phase, Older TLP, has a much wider distribution. Outside the Carpathian basin, it occurs in Germany, Austria, and according to new research, in Poland (Kulczycka-Leciejewiczowa, 1968). The sites shown on Quitta's map (1960) mostly belong to this phase. The last phase in the Central region of the Linear Pottery culture is characterised by the development of the Notenkopf decoration and later on by the appearance of the Zseliz style. Notenkopf and Zseliz sites in Hungary are mostly found in the north, including both Transdanubia and the foothills of the Northern Mountains in eastern Hungary. It is not unlikely that they represent only a local development, and further south the Older Linear Pottery culture may have survived into the latest phase. This, however, still needs confirmation by further research. Radiocarbon dates are again few and come only from the older and later phases: they range from 4230 to 3485 bc. The material culture of the Central Linear Pottery area has been described by Tringham (1971).

For the sake of simplicity, the Linear Pottery development of Hungary will be divided in the following study into only three horizon.. The Early phase, which includes the Szatmár group and the Very Early Alföld Linear Pottery in the east, and the Medina group and the Very Early Transdanubian sites in the west; the Middle phase, which comprises the sites of the developed ALP, and the Older TLP; and finally, the Late phase, which includes the various local groups in the Alföld and the Notenkopf and Zseliz groups in Transdanubia. The chronological position of the sites is shown in Appendix A, and the sites are indicated separately in Map 2.

ENVIRONMENTAL ANALYSIS

A. CLIMATE

Hungary, situated between latitudes 45°48' and 48°35' N, lies in the temperate climatic zone. In Köppen's system of world climates, the climate of eastern Hungary (Alföld) is classified as Cfa and the remainder of the country as Cfb (Trewartha, 1954). 'C' designates warm temperate rainy climates, where the average temperature of the coldest month is below 18° but above -3°C and the average temperature of the warmest month is over 10°C. 'f' stands for no distinct dry season; 'a' for hot summers with the average temperature of the warmest month over 22°C; and 'b' for cool summers with the average temperature of the warmest month under 22°C. In Köppen's classification, the similarity of western Hungary's climate to that of western Europe is emphasised as Transdanubia is grouped together with western Europe in one climatic region.

In Trewartha's (1954) modified Köppen system, most of Hungary belongs to the Daf climatic region. 'Da' is defined as humid continental with warm summers. 'f' is the same as in Köppen's: rain in all seasons with accent on summer. While Köppen uses the coldest month isotherm of -3°C as a boundary between C and D (meso- and microthermal) climates, Trewartha puts the boundary at 0°C. D climates as opposed to C climates have more pronounced winters and longer lasting snow cover. The greater severity of the climate is partly due to the fact that D climates occupy more interior and leeward locations on the great landmasses and their climate is landcontrolled to a much greater extent. In Trewartha's classification, the continental character of Hungary is emphasised, and Hungary is classed together with the major part of Yugoslavia, Romania and the north-western half of Bulgaria in one climatic region.

The differences between the two classifications are relatively minor; nonetheless,they serve to call attention to the transitional nature of Hungary's climate. Beside the country's position in the mid-latitudes, the climate of Hungary is influenced by three main factors: 1) the country's relatively great distance (1200 km) from the Atlantic Ocean; 2) its proximity to the Mediterranean and Adriatic Seas (300 and 700 km); and 3) relief. The influence of the first two factors will be apparent from the detailed description of the main climatic elements.

The influence of relief: The continental character of Hungary's climate is partly due to the country's position in the middle of an enclosed basin. Hungary is surrounded on three sides by the mountain ranges of the Alps and the Carpathians. The Alps on the western border often divert or suppress the moisture laden westerly winds or cause them to lose much of their moisture content before reaching the basin (Pécsi and Sárfalvi, 1964). The

Table 1. Monthly and annual totals of sunshine hours: 50-year averages from 1901 to 1950 (From Pécsi and Sárfalvi, 1964)

	I	II	III	IV	V	VI	VII	VIII	IX	X	XI	XII	Total
Magyaróvár	60	82	138	177	249	257	276	259	185	125	64	44	1916
Keszthely	63	94	142	178	242	260	281	265	189	129	71	48	1962
Pécs	57	94	136	178	247	263	297	278	187	141	75	49	2002
Budapest	59	84	136	186	254	271	296	270	193	135	68	39	1991
Kecskemét	60	90	140	186	260	275	300	280	195	145	75	45	2051
Szeged	63	92	140	187	249	266	204	285	202	150	80	50	2068
Nyiregyháza	62	77	136	188	253	260	274	262	188	137	67	46	1950
Kékes Peak*	110	87	129	159	228	238	277	253	186	144	93	78	1982

* 13-year average for 1932-1944

Table 2. Monthly and annual average temperatures in centigrades: 50-year averages (from Bacsó, 1971)

	I	II	III	IV	V	VI	VII	VIII	IX	X	XI	XII	Year
Magyaróvár	-1.6	-0.1	4.9	9.9	15.0	17.9	20.0	19.1	15.4	9.9	4.4	0.4	10
Pécs	-0.7	1.0	6.4	11.4	16.4	19.8	22.0	21.5	17.4	11.5	5.7	1.6	11
Budapest	-1.0	0.9	6.0	11.1	16.4	19.5	21.5	20.8	16.7	11.0	5.1	1.3	10
Kecskemét	-2.0	-0.1	5.3	10.7	16.2	19.3	21.5	20.6	16.3	10.6	4.6	0.2	10
Szeged	-1.2	0.6	6.3	11.4	16.8	20.0	22.3	21.4	17.5	11.9	5.9	1.4	11
Nyiregyháza	-3.0	-1.1	4.5	10.1	15.8	18.7	20.7	19.7	15.6	9.9	4.1	-0.4	9
Kékes	-4.0	-3.3	0.2	4.7	10.4	13.2	15.7	15.1	11.9	6.1	0.7	-2.2	5

Table 3. Monthly and annual sums of precipitation (mm): 50-year averages (from Bacsó, 1971)

	I	II	III	IV	V	VI	VII	VIII	IX	X	XI	XII	Total
Magyaróvár	36	34	38	43	65	57	64	58	51	49	51	48	594
Pécs	38	39	44	63	67	64	58	56	54	66	67	45	661
Budapest	39	39	43	52	69	67	50	48	45	54	61	50	617
Kecskemét	26	29	32	44	55	54	48	44	46	48	50	37	513
Szeged	32	34	38	49	61	68	51	48	47	52	52	41	573
Nyiregyháza	29	30	32	44	61	70	64	68	46	51	50	38	583
Kékes	39	41	47	61	86	98	79	75	61	69	74	54	784

Carpathians, on the other hand, often serve as a moderating influence. They provide some protection against the invasion of cold air masses from the Arctic in winter and against the hot and dry east wind from Russia and Romania in summer. Temperature conditions are in consequence less extreme than further east. However, cold dry air masses trapped by the mountain chain may settle over the basin for considerable periods of time.

Within Hungary, some of the local climatic differences are also due to relief, although on the average only a change of 0.5°C can be attributed to orographic differences (Bacsó, 1971). Local microclimatic effects are, however, important, especially on south facing slopes. The amounts of heat absorbed by these slopes approximate those measured in the plains of north Sicily and southern Spain (Stefanovits, 1971).

Solar radiation: Due to its more continental climate, Hungary receives much more abundant solar radiation than the countries of the same latitudes in the west. The annual total of sunshine hours varies between 2050 and 1700. As shown in Table 1, the maximum amount of sunshine is received by the area between the Danube and the Tisza, while the poorest in sunshine are the western parts of the country. The abundance of solar radiation, especially in the summer half-year, is favourable for the production of crops with high sunshine requirements. Originally mediterranean plants, the vine, fig, chestnut, quince, lavendar, peach and almond grow well in Hungary (Stefanovits, 1971).

Temperature: D type climates have relatively severe seasons with large variations in annual ranges. The annual absolute temperature variation in Hungary is 70°C in general, but can be as high as 75°C (Pécsi and Sárfalvi, 1964) 1964). The annual mean temperatures vary according to geographical location in the country from below 8° to above 11°C. As seen in Table 2, the southern part of the country is on the average 3°C warmer than the north.

The coldest month is January (0-4°C), and the coldest part of the country is the northeastern region. The warmest month is July (10-23°C), and the hottest part is the southeast. The long term mean annual variation of monthly means is greatest in the Alföld (25°C).

An important figure for agriculture, the length of the growing season varies between more than 210 and less than 160 days per year. The frost free period is longest in the south and gradually decreases toward the north. In comparison with western Europe, Hungary has a higher annual mean temperature and, especially in the Alföld, also a higher summer mean temperature (Sielmann, 1972).

Precipitation: In D climates, summer is normally the season of maximum precipitation. The pressure system and windbelt of both summer and winter are characterised by the dominance of westerlies. In winter, however, continental anticyclones, which develop over the colder parts of the landmasses, tend to repel or divert the cyclonic (moisture-bearing) storms. In summer, although the cyclones are fewer and weaker, they can penetrate deeper into the continents (Trewartha, 1954). Additional reasons for higher precipitation in summer are: 1) the specific humidity of water vapour in the atmosphere is much less over the continent during the cold winter; 2) in winter the subsiding air in the continental anticyclone is likewise conducive to low specific humidity;

3) convection is at a maximum in the summer; and 4) a tendency toward a monsoon system of winds develops due to seasonal extremes of temperatures and pressures.

In the more subhumid interior locations, the period of maximum rainfall is early summer and late spring. According to Trewartha, this is due to the rapid warming up of the earth surface under strong insolation, which leads to atmospheric instability and consequently to convectional overturning. In Hungary, the main maximum of precipitation is in May, June and July. There is also a second maximum of precipitation in autumn, which is especially pronounced in the southwestern part of the country.

According to Köppen (Köppen and Geiger, 1932), there is a transitional zone between the mediterranean winter and continental summer rainfall belts. This so-called 'submediterranean or equinoctial' belt is characterised by a double maxima of precipitation, in spring and autumn. The autumn maximum is probably due to the equatorward creep of the cyclone belt, while its poleward movement early in the year may be an additional reason for the spring maximum of rainfall at lower latitudes.

Zólyomi (1964) has analysed the precipitation data of Budapest and found that 27% of the 80 years for which records were kept had a 'submediterranean' character, while the same figure for Transdanubia, especially in the southwest, was 61%. Midsummer in the Pannonian basin is as dry as midwinter.

Due to greater distance to the Atlantic plus the rainshadow effect of the Alps, precipitation in Hungary is much more irregular and infrequent than in western Europe. The annual precipitation varies between less than 500 and more than 900 mm. Regional variations are considerable, and the most noticeable feature is the steady decrease of precipitation from west to east (Table 3). The precipitation is highest along the slopes of the Alps on the western frontier (1000 mm). In the Alföld, annual means are lower than 600 mm, and in the central parts of the plain the mean is below 480 mm. In the hilly and mountainous regions of Transdanubia and northern Hungary, the precipitation is between 600-800 mm.

The number of snowy days is 30-60. The thickness of the snow varies between 4 and 10 cm (Péczely, 1967). In some years, the Alföld, especially its southeastern part, may be completely without snow cover all through the year.

Evapotranspiration: Evapotranspiration, an important factor for agriculture, is a function of precipitation and temperature conditions. Potential evapotranspiration is greatest in the southeast (700 mm) and gradually decreases toward the north and west (580 mm) (National Atlas of Hungary, 1967). Yearly moisture deficiencies vary between 175 and 25 mm. It is the greatest in the central part of the Alföld. The amount of excess water varies between zero (Alföld) and 100-150 mm in the western part of Transdanubia (Péczely, 1967). Droughts in the Alföld show a fair degree of periodicity; they occur on the average every 3 to 5 years when annual precipitation can be as low as 400-500 mm. Years with an excess of precipitation (700-900 mm) are equally frequent (Pécsi and Sárfalvi, 1964).

The composite effect of the various climatic elements have been summarised in the 'Climatic Regions' map of Hungary (National Atlas of Hungary,

MAP 3. Settlement Distributions and Climatic Regions

Key

A. Warm region
(continental influence preponderant)

A1	dry (severe drought), hot summer insufficient humidity over entire year
A2	dry (severe drought), moderately hot summer, insufficient humidity over entire year
A3	moderately dry, hot summer, insufficient humidity in growing season
A4	moderately dry, moderately hot summer, insufficient humidity in growing season
A5	moderately dry, cold winter, insufficient humidity in growing season
A6	moderately humid, mild winter

B. Warm to moderate region
(oceanic influence preponderant)

B1	dry with mild winters, insufficient humidity over entire year
B2	dry with cold winters, insufficient humidity over entire year
B3	moderately dry with mild winters, insufficient humidity in growing season
B4	moderately dry with cold winters, insufficient humidity in growing season
B5	level region, mild winters with frequent fog, moderately humid
B6	hummocky region with mild winters, moderately humid
B7	humid region with mild winters
B8	cold-winter region, moderately humid

C. Cool region
(mountain climate)

C1	moderately humid mild winters
C2	moderately humid, cold winters
C3	humid, mild winters
C4	humid, cold winters region is more than 500 m above sea level

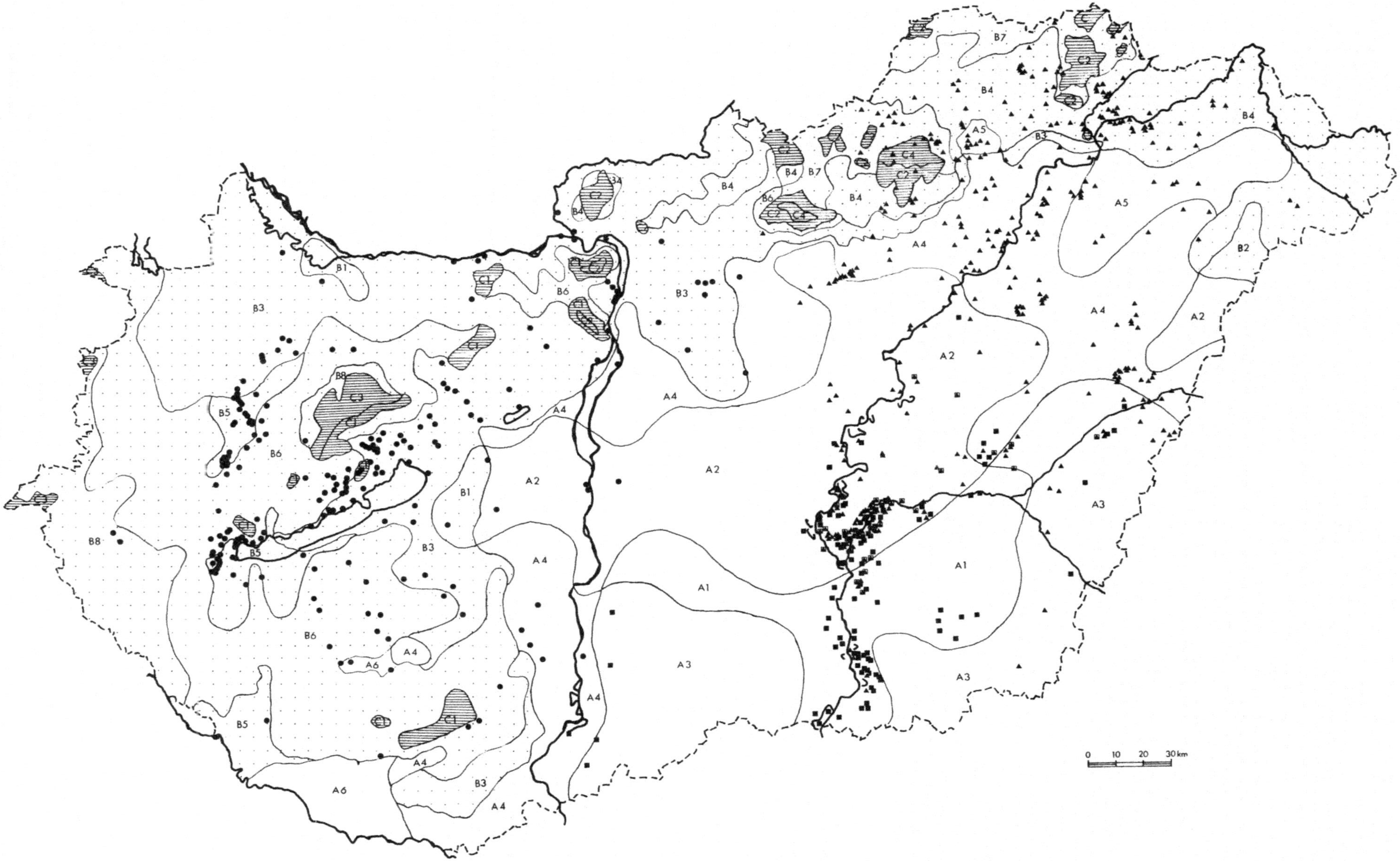
0 10 20 30 km

Fig. 1 Histograms showing the distribution of settlements in relation to climatic regions.

A) Warm region: continental influence preponderant

A1, insufficient humidity over entire year, dry (severe drought), hot summers; A2, insufficient humidity over entire year, dry (severe drought), moderately hot summers; A3, insufficient humidity in growing season, moderately dry, hot summers; A4, insufficient humidity in growing season, moderately dry, moderately hot summers; A5, insufficient humidity in growing season, moderately dry, cold winters; A6, moderately humid, mild winters.

B) Warm to moderate region: oceanic influence preponderant

B1, insufficient humidity over entire year, dry, mild winters; B2, insufficient humidity over entire year, dry, cold winters; B3, insufficient humidity in growing season, moderately dry, mild winters; B4, insufficient humidity in growing season, moderately dry, cold winters; B5, moderately humid, level region, mild winters with frequent fog; B6, moderately humid, hilly region with mild winters; B7, moderately humid, cold winters; B8, humid region, mild winters.

C) Cool region: mountain climate

C1, moderately humid, mild winters; C2, moderately humid, cold winters; C3, humid, mild winters; C4, humid, cold winters —region above 500 m sea level.

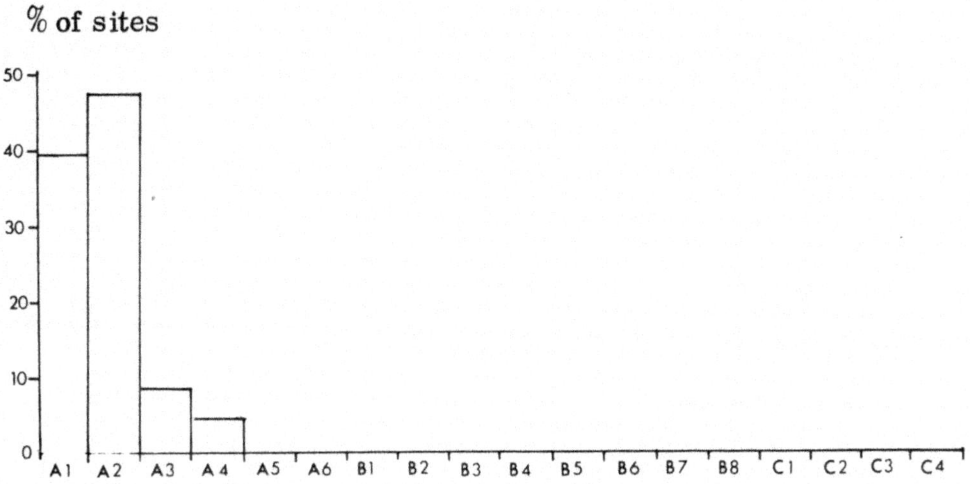

a) Körös culture

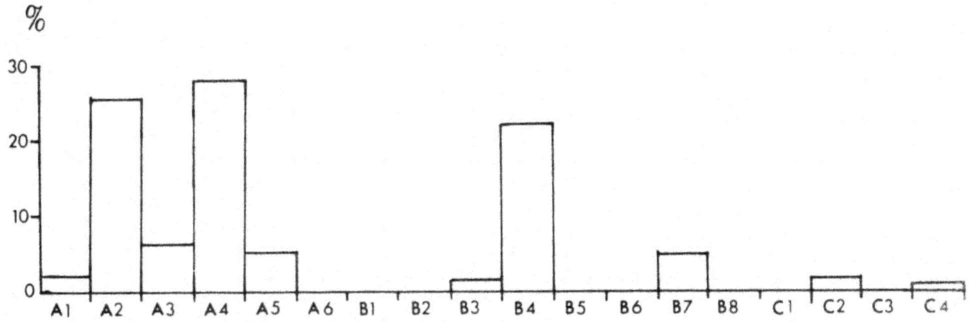

b) Alföld Linear Pottery culture

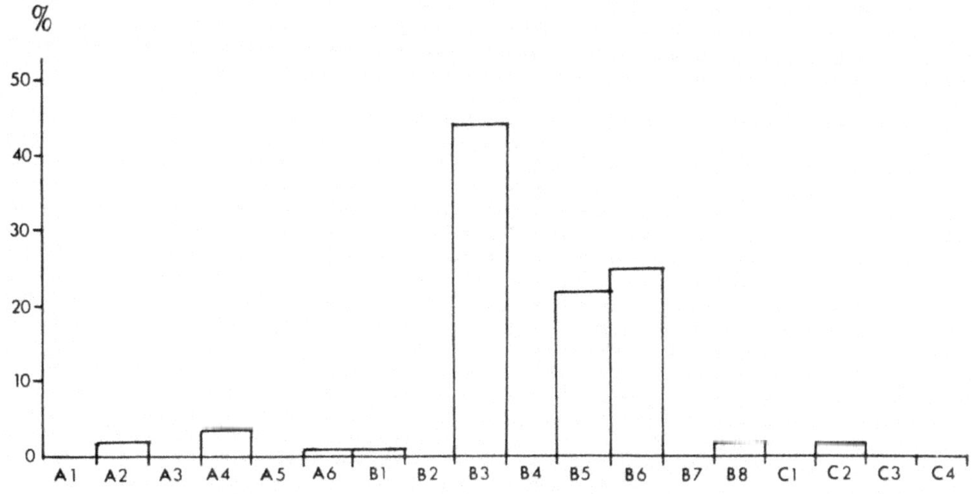

c) Transdanubian Linear Pottery culture

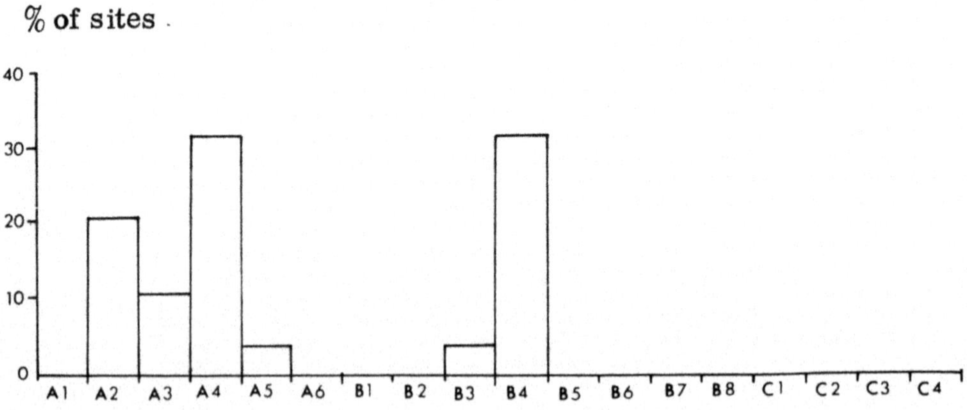

d) Early ALP

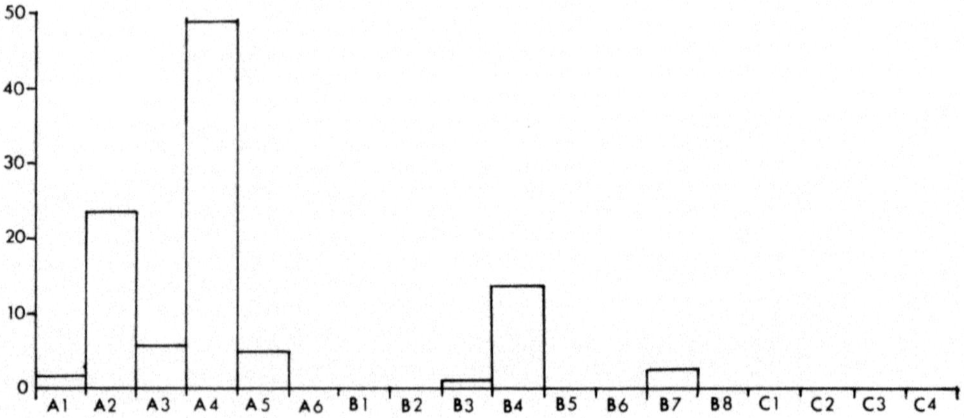

e) Middle ALP

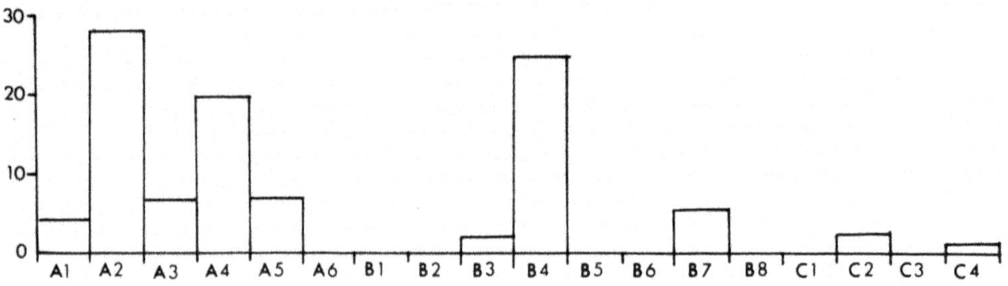

f) Late ALP

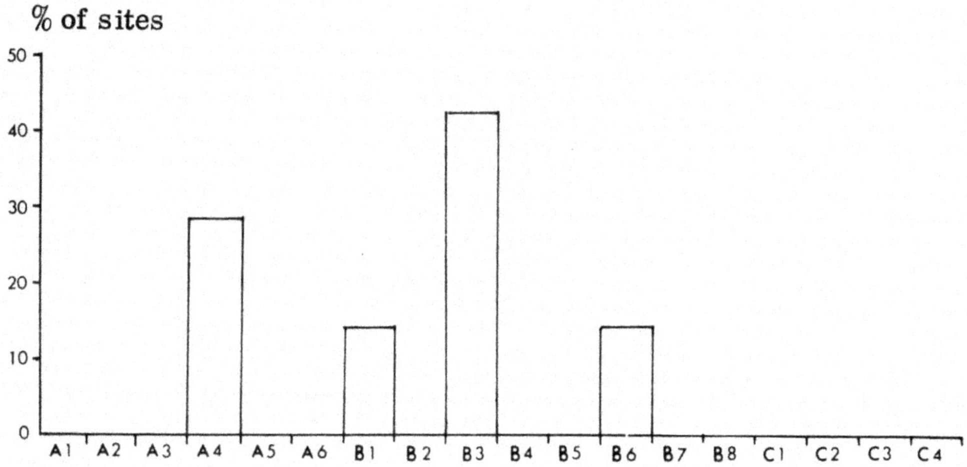

h) Early TLP

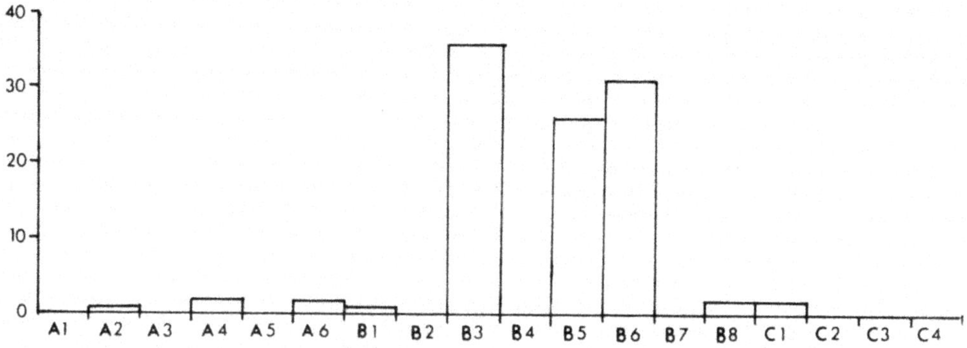

i) Middle TLP

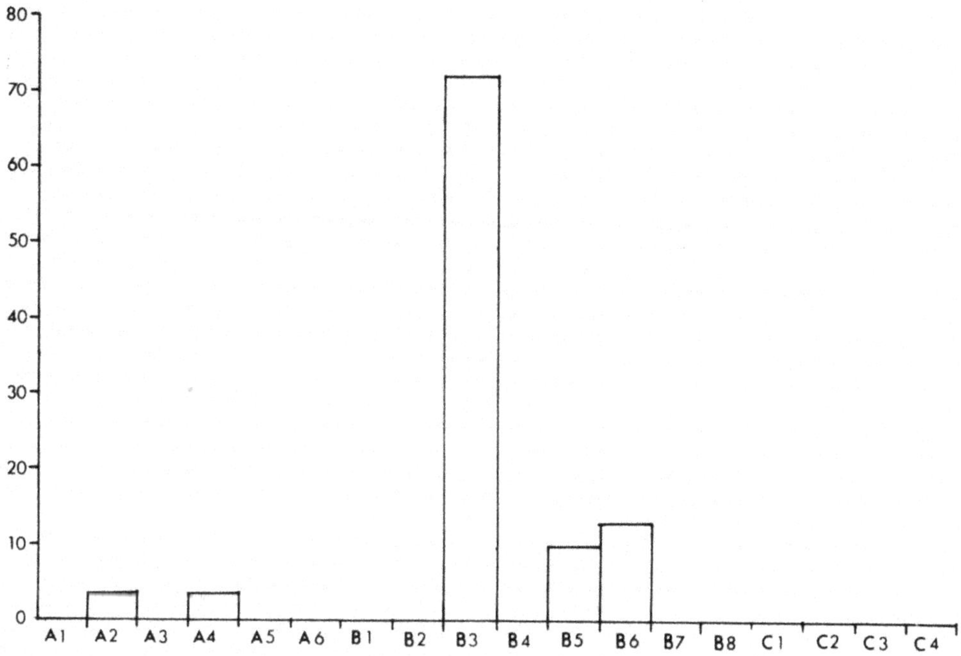

j) Late TLP

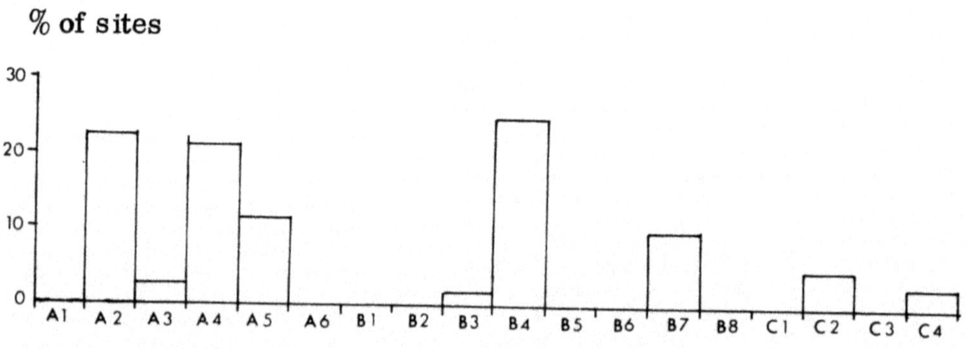

k) Late AVK

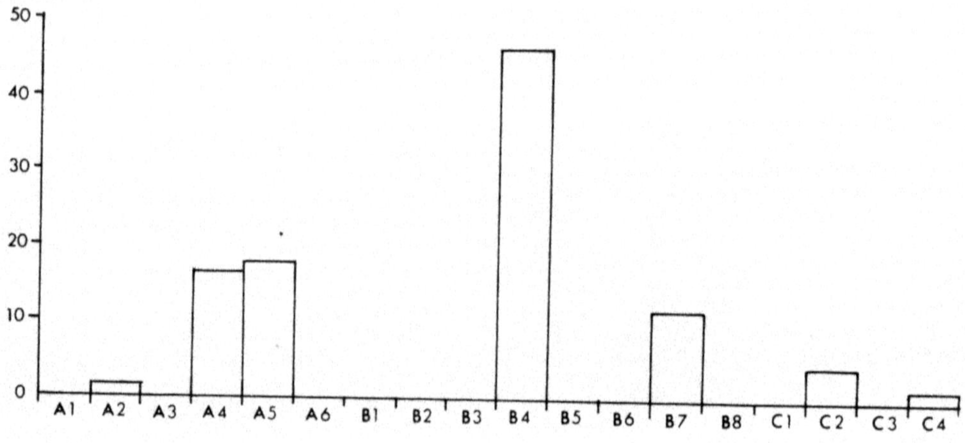

l) Tiszadob

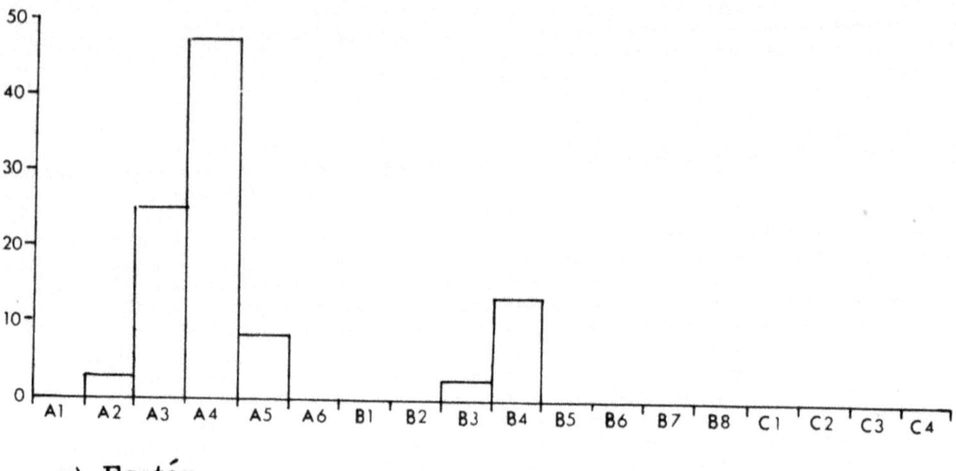

m) Esztár

% of sites

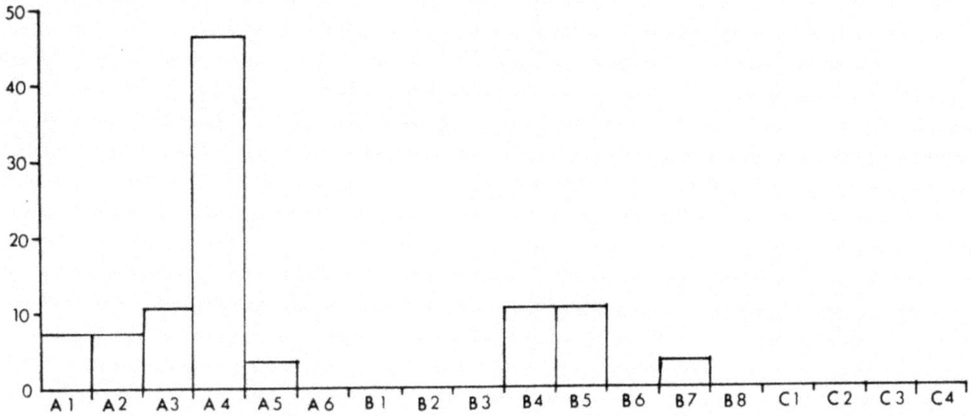

n) Szilmeg

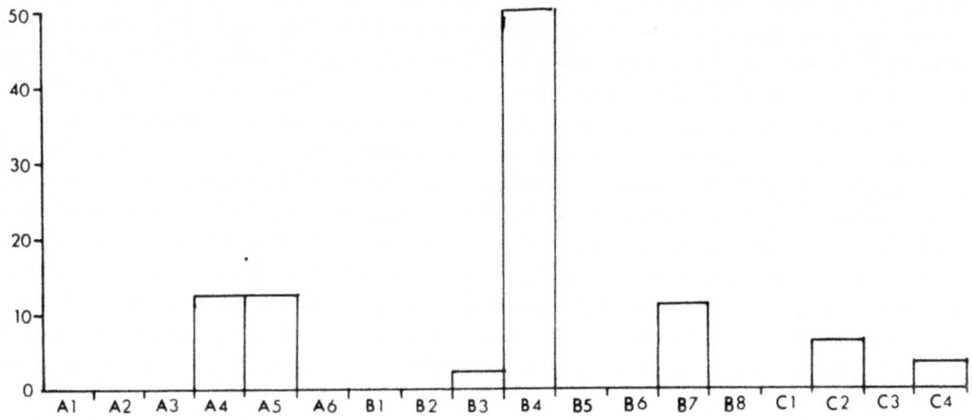

o) Bükk

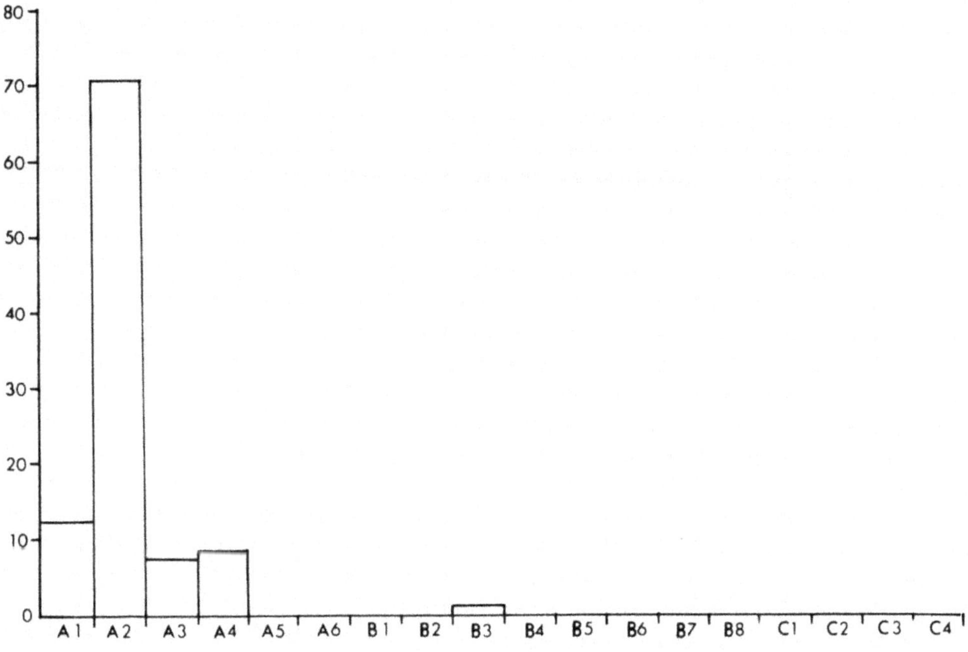

p) Szakálhát

Table 4. Percentage point difference between expected and observed distributions of settlements in relation to climatic regions.

Climate zones	Körös culture Expected %	Körös culture Observed %	Körös culture Ppd	ALP Observed %	ALP Ppd	TLP Observed %	TLP Ppd
A1	5.76	39.69	-33.93	2.86			
A2	14.64	47.33	-32.69	25.97	-11.33	1.89	
A3	10.65	8.40		6.49			
A4	13.88	4.58		28.57	-14.69	3.30	
A5	2.39			5.19	- 2.80		
A6	1.25					0.94	
B1	0.79					0.94	- 0.15
B2	0.43						
B3	16.90			1.56		43.87	-26.97
B4	8.04			22.08	-14.04		
B5	2.50					21.70	-19.20
B6	14.24					24.53	-10.29
B7	2.15			4.94	- 2.79		
B8	3.21					1.42	
C1	0.92						
C2	1.47			1.56	- 0.09	1.41	
C3	0.54						
C4	0.24			0.78	- 0.54		
Total Ppd			66.62		46.28		56.61

Climate zones	ALP Early Observed %	ALP Early Ppd	ALP Middle Observed %	ALP Middle Ppd	ALP Late Observed %	ALP Late Ppd
A1			1.59		3.97	
A2	20.69	- 6.05	23.02	- 8.38	27.78	-13.14
A3	10.35		5.56		6.35	
A4	31.03	-17.15	48.41	-34.53	19.44	- 5.56
A5	3.45	- 1.06	4.76	- 2.37	6.75	- 4.36
B3	3.45		0.79		1.98	
B4	31.03	-22.99	13.49	- 5.45	24.60	-16.56
B7			2.38	- 0.23	5.56	- 3.42
C2					2.38	- 0.91
C4					1.19	- 0.95
Total Ppd		47.25		50.96		44.89

PPd = Percentage point difference

Table 4 (contd.)

Climate zones	TLP Early Observed %	Ppd	TLP Middle Observed %	Ppd	TLP Late Observed %	Ppd
A2			0.63		3.17	
A4	28.58	-14.70	1.89		3.17	
A6			1.26	- 0.01		
B1	14.28	-13.49	0.63			
B3	42.86	-25.96	35.22	-18.32	71.43	-54.53
B5			25.78	-23.28	9.53	- 7.03
B6	14.28	- 0.04	30.81	-16.57	12.70	
B8			1.89			
C1			1.89	- 0.97		
Total Ppd		54.19		59.15		61.56

Climate zones	Late AVK Observed %	Ppd	Late ALP Tiszadob Observed %	Ppd	Late ALP Esztár Observed %	Ppd
A2	22.22	- 7.58	1.49		2.78	
A3	2.47				25.00	-14.35
A4	20.99	- 7.11	16.42	- 2.54	47.22	-33.34
A5	11.11	- 8.72	17.91	-15.52	8.33	- 5.94
B3	1.23				2.78	
B4	24.69	-16.65	46.27	-38.23	13.89	- 5.85
B7	9.88	- 7.73	11.94	- 9.79		
C2	4.94	- 3.47	4.48	- 3.01		
C4	2.47	- 2.23	1.49	- 1.25		
Total Ppd		53.49		70.34		59.48

Climate zones	Late ALP Szilmeg Observed %	Ppd	Late ALP Bükk Observed %	Ppd	Late ALP Szakálhát Observed %	Ppd
A1	7.15	- 1.39			12.19	- 6.43
A A2	7.15				70.73	-56.09
A3	10.71	- 0.06			7.32	
A4	46.43	-32.55	12.82		8.54	
A5	3.57	- 1.18	12.82	-10.43		
B3			2.56		1.22	
B4	10.71	- 2.67	50.00	-41.96		
B5	10.71	- 8.21				
B7	3.57	- 1.42	11.54	- 9.39		
C2			6.41	- 4.94		
C4			3.85	- 3.61		
Total Ppd		47.48		70.33		62.52

1967). The map is based on Kakas's (1960) work. Its merit for the locational analysis of the settlements is that instead of giving absolute values for temperature and precipitation, it emphasises the interaction of the main climatic elements.

Settlement Distributions and Climatic Regions

The distribution of settlements is shown in Map 3. Percentage values for the number of settlements in each climatic region have been calculated for the three cultures and are shown in Histograms a, b, and c (Fig. 1).

In the Körös culture, 87% of the settlements are in areas where humidity is insufficient all through the year. The remaining 13% are in areas where moisture is deficient in the growing season. All the settlements are in the warm region. In the ALP, 29% of the settlements are in areas where humidity is insufficient over the entire year, and 64% are in areas where moisture is deficient in the growing season. Altogether 93% of the settlements are located in areas which have a negative water balance. 36% of the sites are located in cold winter regions.

In the TLP, less than 3% of the sites are in areas with insufficient humidity over the entire year, and 47% in areas with moisture deficiency in the growing season. Altogether 50% of the sites are in areas with sufficient moisture through the year. Only 1% of the sites are in the cold winter area.

Histograms d, e, f, h, i and j (Fig. 1) show the percentage relationships of settlements to climatic regions in the Early, Middle and Late phases of the ALP and TLP.

The percentage of settlements which are in areas of insufficient humidity over the entire year is 21 in the Early, 25 in the Middle, and 32 in the Late ALP. The percentage for settlements in areas of insufficient humidity in the growing season is 79 in the Early, 73 in the Middle, and 59 in the Late ALP. Altogether the percentage of settlements which are in areas of moisture deficiency is 100, 98 and 91 in the Early, Middle, and Late ALP respectively.

The percentage of settlements in areas of insufficient humidity over the entire year is 14 in the Early, 1 in the Middle, and 3 in the Late TLP. Percentages for the sites in areas of insufficient humidity in the growing season are 71 in the Early, 37 in the Middle, and 71 in the Late TLP. Altogether the percentages for settlements which are in areas of deficient moisture are 85, 38 and 74 in the Early, Middle and Late phases, respectively. All the settlements in all the phases are in mild winter areas.

Percentage figures calculated for the regional groups of the Late ALP are shown in Histograms k, l, m, n, o and p (Fig. 1). The percentage of sites with insufficient humidity over the entire year are: 22 in the Late AVK; 1 in the Tiszadob; 3 in the Esztár; 14 in the Szilmeg; O in the Bükk; and 83 in the Szakálhát group. The percentages of sites with insufficient humidity in the growing season are: 60 in the Late AVK; 81 in the Tiszadob; 77 on the Esztár; 71 in the Szilmeg; 78 in the Bükk; and 17 in the Szakálhát group.

Chi-square values calculated for the Körös, ALP and TLP cultures are significant at $\alpha = 0.005$ level, suggesting that it is extremely unlikely that the distributions are due to chance. Percentage point difference calculated to measure the influence of climate type as a variable on settlement distribution is quite high in all three cultures: 66.62, 46.28 and 56.61 (Table 4). (In the tables, settlement percentages are shown to two decimal points.) Some of the relatively low values (less than 50) in the ALP can be attributed to the influence of regional and chronological differences which are quite pronounced in the ALP.

The most striking feature of the observed settlement distribution is the high percentage of sites in all three cultures which are found in moisture deficient areas. It has been suggested by Sielmann (1917), among others, that as the wild prototypes of the early domesticates prefer a relatively dry habitat in their original homeland, selection for the drier lands in Europe could be viewed as a form of difference minimising strategy. The early farmers by opting for lands least different from the original habitat would have to some extent reduced the impact of changing climatic conditions.

It should be pointed out, however, that the range of precipitation across the belt of wild cereals is much greater (400-1000 mm) (Zohary, 1969) than that suggested by Sielmann (500-600 mm), and consequently some strains more used to greater precipitation could have existed already in the original habitat. Nevertheless, drought must have been a problem in these areas. The homeland of the wild cereals is in the winter rainfall belt, where summers are very warm and dry. The wild progenitors of wheat and barley escaped rather than endured the drought by maturing early.

The maturation time of autumn sown cereals in Hungary is early summer as opposed to the Near East where it is usually early spring. For the growth of annual crops without irrigation, the ratio of rainfall and open-water evep-oration must remain close to 1.00 (Flannery, et al., 1971). As we have seen, summers are dry in Hungary, and most of the country, especially the eastern part, has a negative water budget. The harmful effect of inadequate moisture on crop production is alleviated somewhat by having the rainfall maximum early in the summer, before evaporation reaches its maximum in mid-summer. In consequence, more moisture is available for the cereals in their tillering period, which is also the time of greatest moisture requirement (Trewartha, 1954). Nonetheless, precipitation in Hungary, especially in the Alföld, is extremely unpredictable. In the southeastern part of the Alföld and in the Tisza region, the probability is that precipitation in 75% of the years will not be enough for most plants to produce even mediocre yields (Enyedi and Szabó, 1955). To ensure reasonable crops every year, irrigation in the Alföld, especially east of the Tisza, is a necessity today (Pécsi and Somogyi, 1971). Even with large scale irrigation, the production of spring sown cereals is limited, no more than 2-3% of the total output (Centre d'Etude de l'Azote, 1960).

The high percentage of early agricultural settlements in the moisture deficient areas of Hungary was probably due to factors other than purely climatic ones, such as nearness to rivers, favourable soil types, etc. Settling in areas of insufficient moisture would have nonetheless effectively limited successful crop production unless some supplementary source of moisture could be assured. As will be shown later, settlements in the Alföld, where the degree of moisture deficiency would have been the greatest, were located in moister habitats within the zonal forest steppe. This is especially true for the Körös and Szakálhát settlements, which occupy the driest regions in the Alföld. The number of settlements in areas of insufficient humidity all through the year decreases significantly in the ALP and is further reduced in the TLP. The amplitude of water deficiency measured in mm is also less for a large percentage of the ALP and TLP sites. Furthermore, 50% of the TLP sites are in areas which are adequately moist. The same figure in the ALP is only 7%.

It is interesting to note that the percentage of sites in areas with insufficient moisture is almost twice as much in the LAte TLP as in the Middle phase. Quitta (1969) suggests a similar pattern in the Linear Pottery areas of Czechoslovakia and Germany. At this point, it is hard to decide whether this pattern is a real one or not. The chronology of the TLP sites, at least in Hungary, is somewhat insecure. It is also possible, as Quitta has suggested, that many of the early sites are buried under later alluvium and are as yet undetected.

As summer drought can be limiting for late maturing plants, so can winter frost be limiting for autumn sown cereals. In Hungary, the crops are usually protected by snow, although frost disasters do occur occasionally. They are most frequent in the Alföld, where the ground is sometimes left without a snowcover. These occurrences are not, however, sufficiently numerous or severe enough to seriously limit the production of winter cereals. A greater problem for mixed farmers in the temperate zone would have been the scarcity of winter fodder for domesticated animals. The effect of this limitation will be discussed later. While cold winters are characteristic of Hungary as a whole, the severity of the winter varies from one part of the country to the other. It is of interest that, while almost all the TLP settlements are in relatively mild winter areas, at least 30% of the ALP sites are located in areas of really cold winters. Transhumance is one possible explanation, which could account for the different patterns.

Climate in the Atlantic Period

Agricultural dispersal in Europe took place in the Atlantic climatic phase (*c*. 6200-3300 B.C.). The above analysis has been made on the implicit assumption that modern conditions are similar to those that prevailed during the time of the first agriculturalists. How far is this assumption justified?

The climate of any one region is determined by two essentially different factors: the prevalent atmospheric circulation pattern on the one hand, and relief and the distribution of land and sea on the other (Butzer, 1961). It is unlikely that the second factor would have been different in the Atlantic phase. Changes in the atmospheric circulation patterns between *c*. 8000-0 B.C. have been tentatively reconstructed by Lamb *et al*. (1966).

The reconstruction is based on surface temperature data, which in their turn are based on paleobotanical evidence. The distribution of certain warmth loving plants suggests that summer temperatures in mid-altitude Europe were at least 2°C higher during the Atlantic (Butzer, 1971). According to Járai-

Komlódi's (1969) calculations, the figures would have been similar in Hungary. Neolithic finds of Asiatic buffalo remains in northern Yugoslavia would also indicate a climate warmer than today's (Bökönyi, 1957).

Examining historical rainfall patterns, Lamb observes that in the western half of Eurasia, where westerly winds prevail 'mild epochs tend to go with high prevalence of southwesterly winds and therefore high sea temperatures, resulting in more moisture input into the atmosphere and moister winds'. As the prevalence of westerly winds in midlatitude Europe dates back to the end of the Tertiary period (Schwarzbach, 1961), it is reasonable to assume that the same relationship would have been true also for the Atlantic period, where the palynological evidence is that of a mild but moist epoch (Firbas, 1949-52).

An interesting observation about the medieval warm epoch (1150-1300 A.D.) is that although it had a higher annual rainfall than today's, the 'high summer months appeared to have been drier than now' (Lamb, et al., 1966) due to the prevalence of anticyclonic tendencies. In his reconstruction of the world circulation patterns, Lamb sees a similar anticyclonic tendency in Europe during the Atlantic phase. Evidence for former river and lake levels would at least partially support the argument. River levels seem to have been similar to today's or even lower during the Atlantic in some parts of Europe. Sulimirski (1961) argues the case for Russia, Vadász (1969) for Hungary and Quitta (1969) for Germany in certain periods of the Atlantic. Butzer (1971) quotes Aario, according to whom increased evaporation reduced stream discharge by at least 30% in northern Europe. According to Sági's (1968) calculations, the level of the Lake Balaton in Hungary was no higher in the Atlantic than it is today. River and lake levels, should, on the other hand, be treated with caution as climate indicators, as variables other than evaporation and rainfall can also affect their fluctuation.

As to differences between western and eastern Europe, Lamb, on the basis of a preliminary analysis for the area of the Russian plain between Kiev and Leningrad, suggests that during the Atlantic average yearly rainfalls were about 90% and 120% of the modern values in Russia and western Europe respectively. This would agree with the more continental character of eastern Europe, where winter precipitation is relatively low in any case and as suggested by Lamb summer rainfall would also have been reduced in the Atlantic.

In summary, the Atlantic period was warmer than now and with a higher annual rainfall average due to the increased sweep of the westerlies. But the amount of rainfall would have shown a decrease toward the east, and it is also likely that summers would have been drier in Europe than today due to increased anti-cyclonic activity in that season.

For Hungary, increased anticyclonic activity in the summer coupled with the country's relatively great distance from the Atlantic would have meant warm dry summers, possibly with frequent droughts, especially in the eastern part of the country. As, in addition, some of the climatic dryness of the country is due to the rain shadow effect of the relief, it is reasonable to assume that insufficient humidity was already a climatic factor in some parts of Hungary in the Atlantic. On the basis of wild ass bones found in Körös and Linear Pottery settlements, Bökönyi (1954; 1957a; Makkay, 1970) has suggested the existence of a relatively dry steppe in Hungary and northern Yugoslavia in the neighbourhood of the settlements during the neolithic. The presence of wild ass at Lepenski Vir is exceptional, but the gorge may have been used as a winter watering place, or the ass could have been brought in for other reasons. Bustard bones at Nosa and Ludas-Budźak also suggest the existence of relatively open country in the neighbourhood (Trogmayer, 1968). Other evidence for the existence of forest steppe in Hungary will be discussed in later chapters.

It is possible that winter cold was less of a problem in Hungary during the Atlantic. But as some of the increase in cold would have been due to orographic factors then, as now, the existence of cold winters cannot be entirely ruled out at least in some parts of Hungary.

Although the climate of the Atlantic period differed in some respects from that of the present era, there is no evidence to suggest that the nature of climatic regions in Hungary or their relative position would have been drastically different. According to Zólyomi's (1958) reconstruction, based on paleobotanical evidence, the Alföld had a Cfaxx'' (s) climate and the south facing slopes of the Hungarian Mountains a Cfax'' (s) climate. (x stands for early spring precipitation maximum; x'' for double maximum of precipitation; and s for summer dryness.) The reconstruction suggests that the climate was essentially similar to that of today's, except the Mediterranean element was perhaps more pronounced in the Atlantic.

In conclusion, it is suggested that the present climate region map of Hungary is acceptable, though with certain reservations, as a rough guide to Atlantic conditions. It should however also be taken into account that some climatic oscillations would have occurred in the course of the Atlantic period (Frenzel, 1966). According to the evidence of pollen diagrams, the middle and the end of the period was drier in Hungary (Borsy, 1961), and Vozáry (1957) distinguishes as many as 14-15 minor local oscillations.

B. VEGETATION

One of the basic principles of plant geography is that climatic control is a primary factor determining the general character of world vegetation (Butzer, 1971). A particular climatic belt will tend to favour a certain association of plants, grasses, shrubs and trees known as a plant formation or as the climatic climax vegetation. 'The basic premise is that if a naturally well-drained surface is left completely undisturbed for a protracted period, with no human activity, climatic change, or other natural cataclysm, a whole series of plant communities, one after another, will occupy it but, ultimately, a community will establish itself and persist, unchanged, quite indefinitely. This climatic climax community will be dominated by plants, which of all those available, can compete most successfully in the existing physical conditions' (Eyre, 1963, 10). The vegetation which is in accordance with the climate zone is called the zonal vegetation.

The vegetation zones of Hungary are chiefly based on information given in the climatic diagrams of Walter's Klimadiagramm Weltatlas (National Atlas of Hungary, 1967). Climate diagrams are graphic representations of the combined and simultaneous effect of the various climatic elements. They show the annual mean temperature and precipitation values at a particular locality, the occurrence, length and intensity of relatively humid and arid seasons, the duration and severity of cold winter and the possibility of late or early frosts (Walter, 1973).

In accordance with its climate, Hungary belongs to two main vegetation zones: the deciduous summer forest zone and the forest steppe (National Atlas, 1967). The deciduous forest here is a climatic climax formation. The ecological status of the forest-steppe is somewhat doubtful, but on vegetation maps it is usually taken as a probable climatic-climax formation (Eyre, 1963). When the climate zonal vegetation map of Hungary is compared with the Climatic Regions map discussed in the previous chapter, it is seen that certain areas which are characterised by insufficient moisture in the growing season, nonetheless support forest growth. This is due to the fact that trees like the oak root deeply in the soil and consequently are able to tap subsoil moisture. The amount of moisture deficiency in these areas is also relatively small.

Settlement Distribution and Vegetation Zones

The distribution of settlements according to vegetation zones is shown on Map 4. Percentage values for the occupation of each zone are shown in Histograms a, b, c (Fig. 2).

The settlements of the Körös culture are all located in the forest steppe zone. The great majority of sites, 96%, is in the dry forest-steppe. The remaining small percentage of sites is found along the margins of the forest steppe. This marginal or transitional belt (zones III and II2) is characterised by less severe dryness. The degree of dryness is calculated as the ratio of duration of dry to wet weather (semiaridity index).

In the ALP, 35% of the sites are in the dry forest steppe, and 37% are located along the less dry margins; 28% of the settlements are in the forest zone. In the TLP, the majority of the sites, 67%, are in the forest zone; 32% in the transitional forest-steppe zone. Less than 2% are located in the dry forest-steppe.

The relationship of settlements to vegetation zones in the Early, Middle and Late Phases of the ALP and TLP are shown in Histograms d, e, f, h, i, and j (Fig. 2).

The percentage of settlements in the dry forest-steppe is 28 in the Early, 29 in the Middle and 39 in the Late phase of the ALP. The percentage of settlements located in the transitional forest-steppe zone is 45 in the Early, 55 in the Middle and 28 in the Late Phase. The percentage figures for the occupation of the forest zone are: 28 in the Early, 16 in the Middle, and 33 in the Late Phase.

The percentage figures for the occupation of the forest zone in the TLP are: 29 in the Early, 77 in the Middle and 41 in the Late Phase. The rest of the sites in the Early and Middle phases, 71 and 23% respectively, are located in the transitional forest steppe zone. In the Late TLP, 55% of the sites are in the transitional zone and 3% in the dry forest-steppe.

MAP 4. Settlement Distributions and Vegetation Zones

Key

I. Zone of Deciduous Forests; Humid Climate

·11·	Zone of montane beech forest: pure beech, and beech with fir
·12·	Zone of submontane beech forests with hornbeam
·13·	Zone of hornbeam oak forests
·14·	Zone of oak forests

II. Zone of Forest-Steppes; Semiarid-Semihumid Climate

111	Forest-steppe in hummocky areas and plain margins
112	Moderately dry forest-steppe on the margin of the Great Plain; short drought period
113	Dry forest-steppe with prolonged drought period
114	Dry forest-steppe with prolonged, severe drought in summer
— 7 —	Semiaridity index (ratio of duration of dry to wet weather)

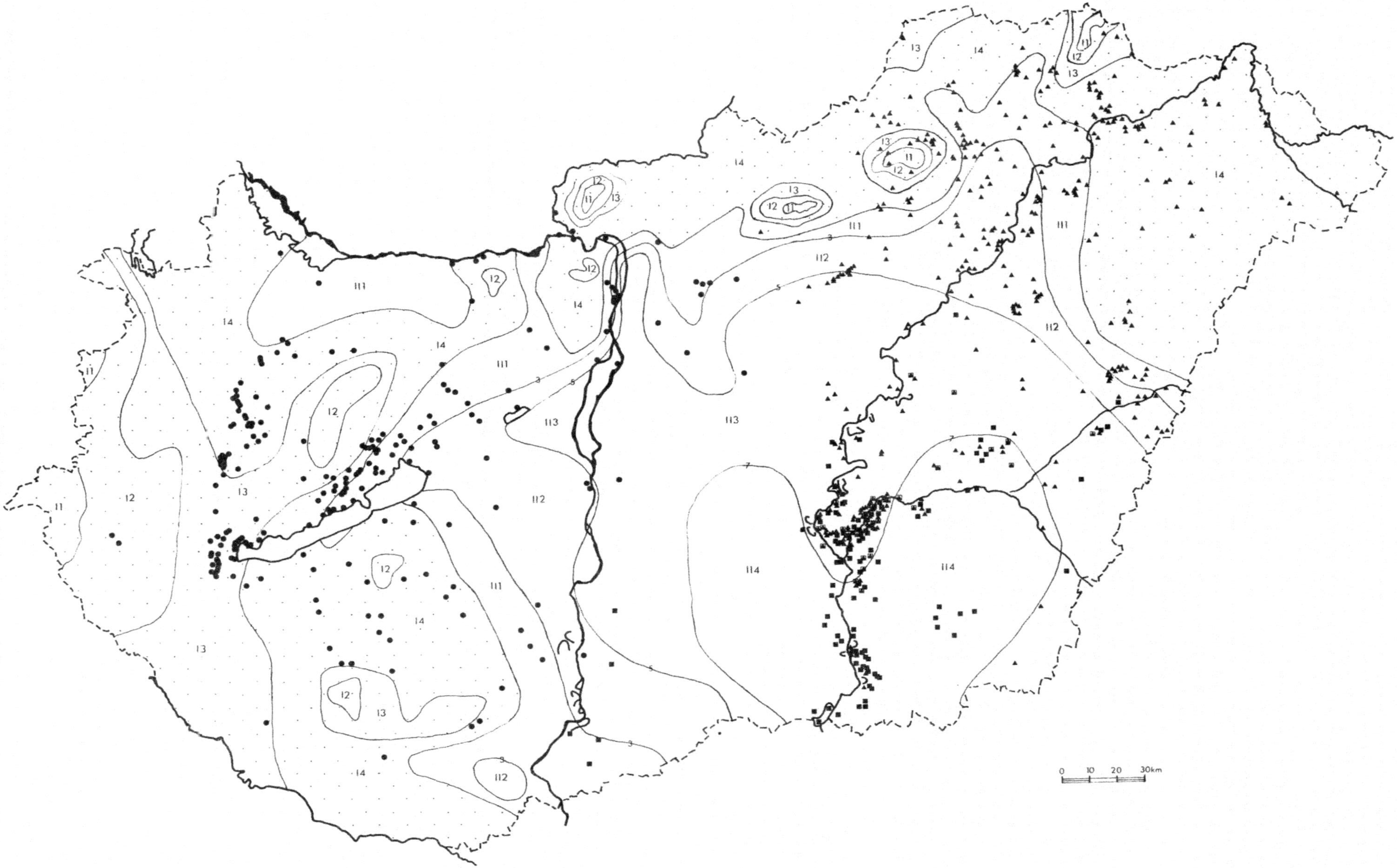
0 10 20 30km

Fig. 2 Histograms showing the distribution of settlements in relation to vegetation zones.

I, Deciduous forest zone

I1, mountain beech forests; I2, low-mountain beech forests with hornbeam; I3, hornbeam-oak forests; I4, oak forests.

II, Forest-steppe zone

II1, forest-steppe in hilly areas, and along the margins of the plains; II2, moderately dry forest-steppe on the margin of the Great Plain with short drought period; II3, dry forest-steppe with prolonged drought; II4, dry forest-steppe with prolonged severe drought.

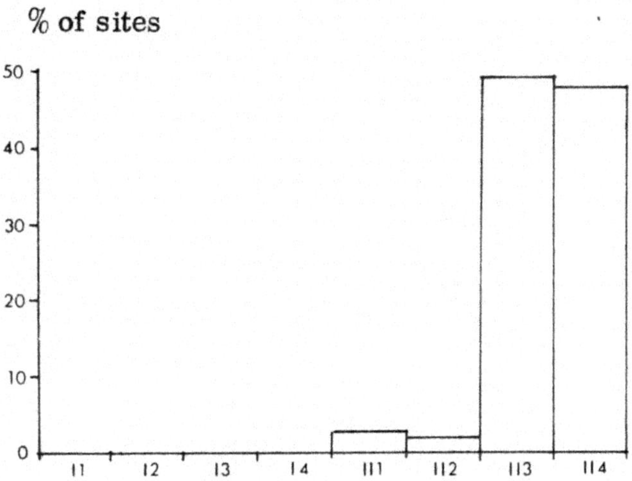

a) Körös culture

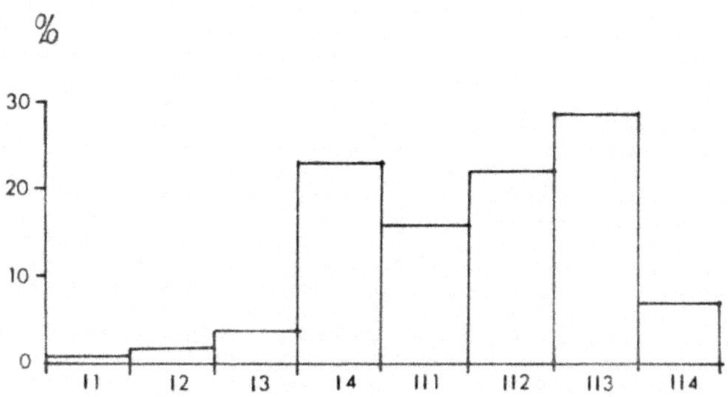

b) Alföld Linear Pottery culture

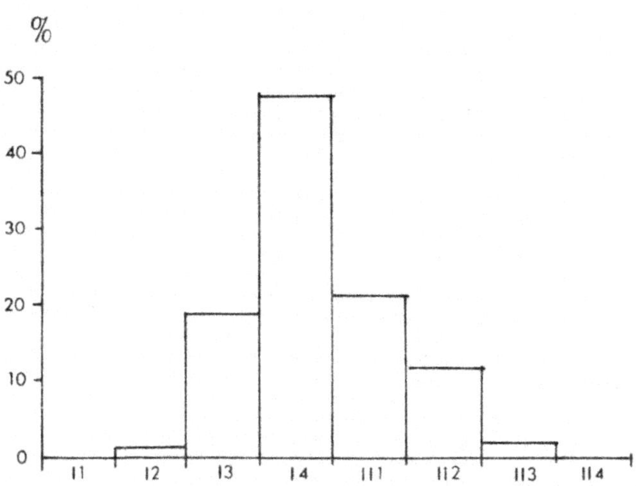

c) Transdanubian Linear Pottery culture

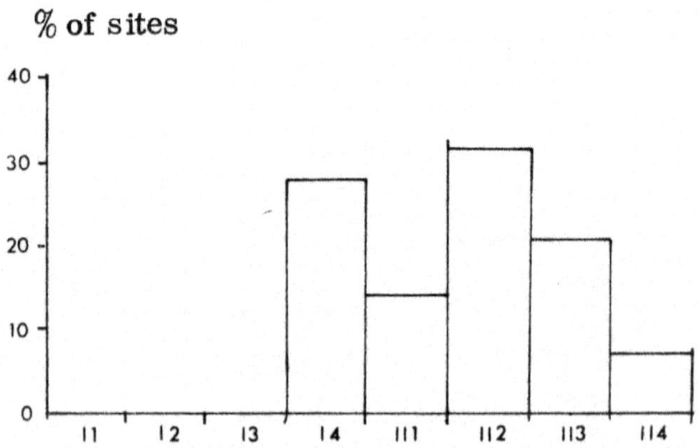

d) Early ALP

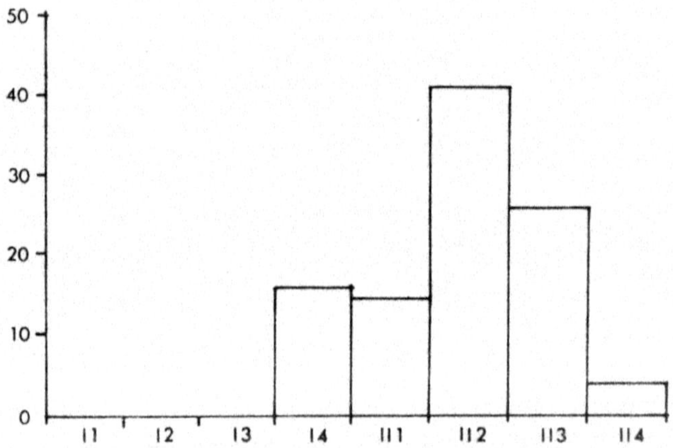

e) Middle ALP

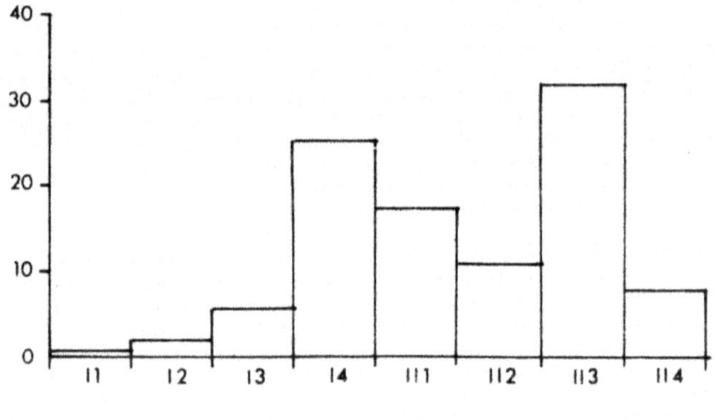

f) Late ALP

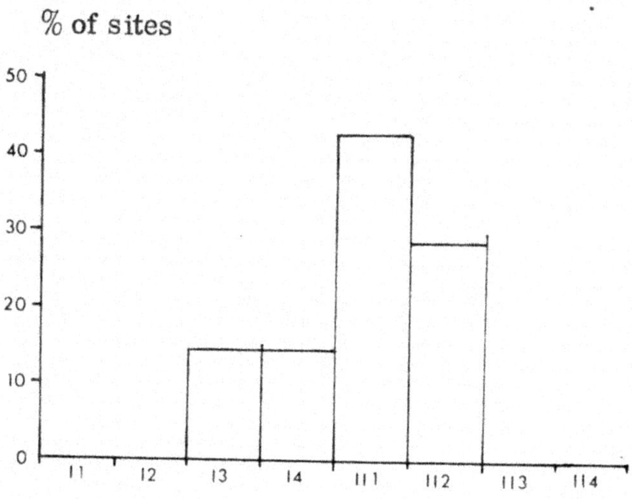

h) Early TLP

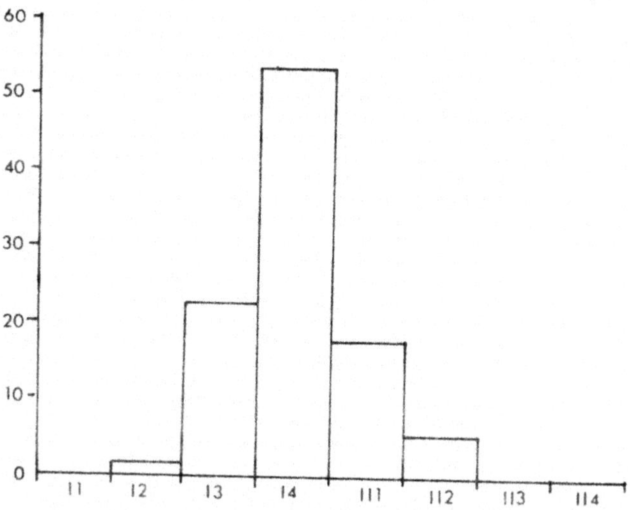

i) Middle TLP

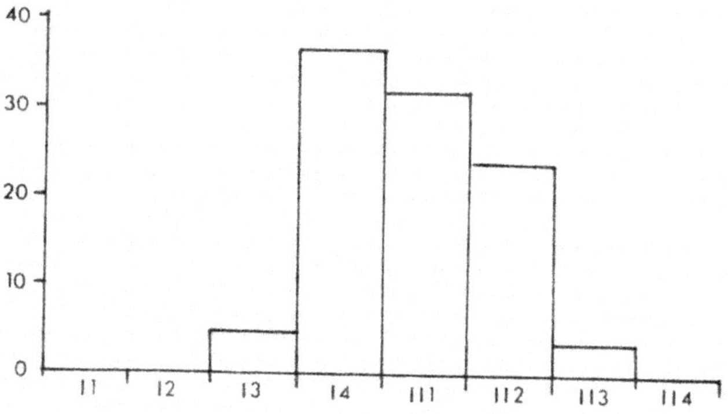

j) Late TLP

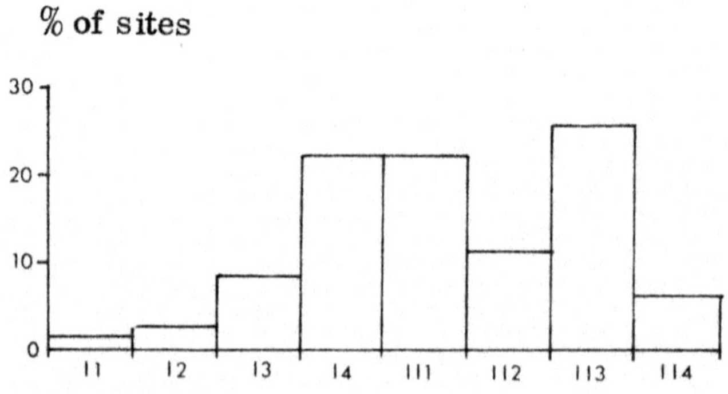

k) Late AVK

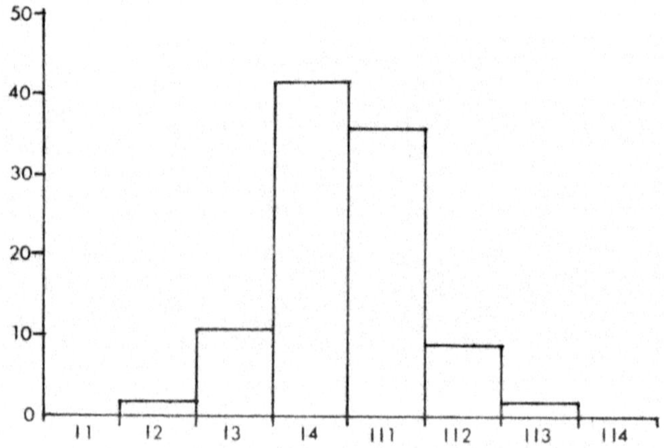

l) Tiszadob

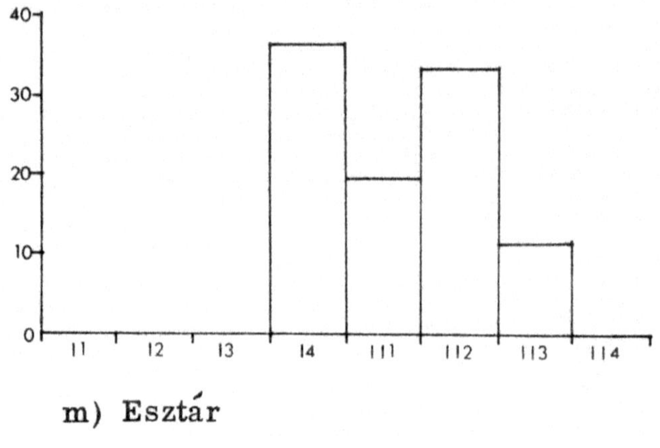

m) Esztár

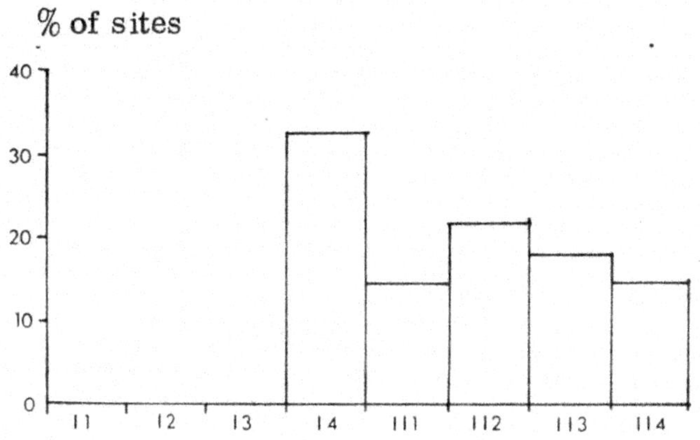

n) Szilmeg

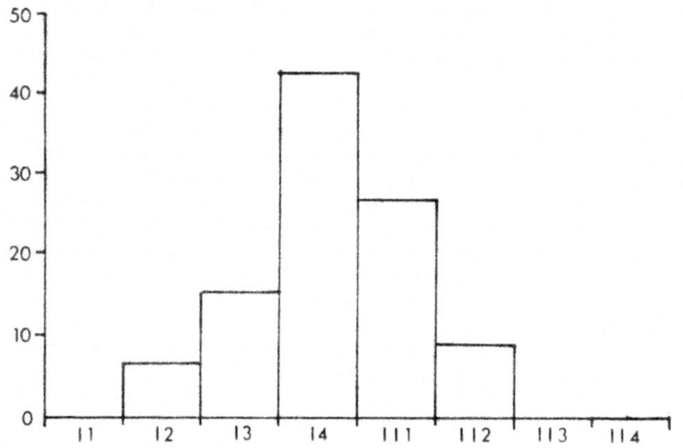

o) Bükk

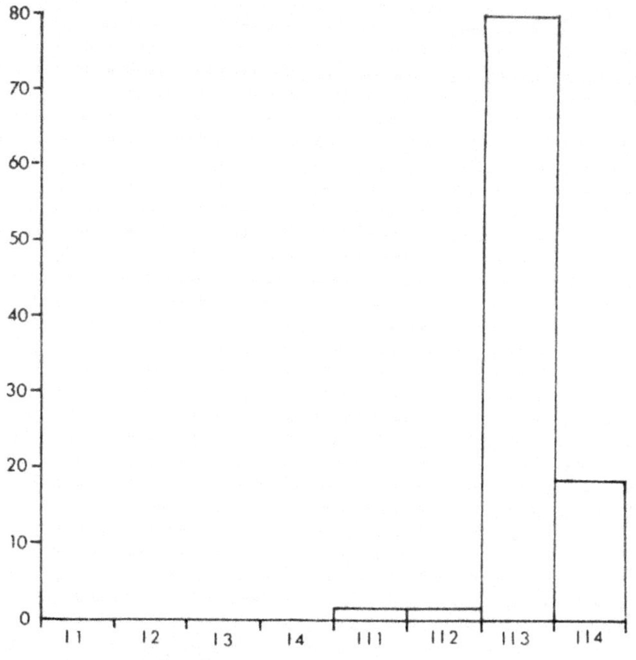

p) Szakálhát

Table 5. Percentage point difference between expected and observed distributions of settlements in relation to vegetation zones.

Vegetation zones	Körös culture Expected %	Körös culture Observed %	Körös culture Ppd	ALP Observed %	ALP Ppd	TLP Observed %	TLP Ppd
I1	1.11			0.26			
I2	4.62			1.30		0.94	
I3	9.05			3.64		18.40	- 9.35
I4	31.49			22.86		47.17	-15.68
II1	11.84	2.29		15.32	- 3.48	20.75	- 8.91
II2	11.60	1.53		21.82	-10.22	11.32	
II3	19.02	48.85	-29.83	28.05	- 9.03	1.42	
II4	11.27	47.33	-36.06	6.75			
Total Ppd			65.89		22.73		33.94

Ppd = Percentage point difference

Vegetation zones	ALP Early Observed %	ALP Early Ppd	ALP Middle Observed %	ALP Middle Ppd	ALP Late Observed %	ALP Late Ppd
I1					0.40	
I2					1.98	
I3					5.56	
I4	27.59		15.87		25.	
II1	13.19	- 1.95	14.28	- 2.44	17.06	- 5.22
II2	31.03	-19.43	40.48	-28.88	10.71	
II3	20.69	- 1.67	25.40	- 6.38	31.75	-12.73
II4	6.90		3.97		7.54	
Total Ppd		23.05		37.70		17.95

	TLP Early Observed %	TLP Early Ppd	TLP Middle Observed %	TLP Middle Ppd	TLP Late Observed %	TLP Late Ppd
I2			1.26			
I3	14.28	- 5.23	22.64	-13.59	4.76	
I4	14.28		53.46	-21.97	36.51	- 5.02
II1	42.86	-31.02	17.61	- 5.77	31.75	-19.92
II2	28.58	-16.98	5.03		23.81	-12.21
II3					3.17	
Total Ppd		53.23		41.33		37.14

(continued)

Table 5 (contd.)

Vegetation zones	Late AVK Observed %	Late AVK Ppd	Late ALP Tiszadob Observed %	Late ALP Tiszadob Ppd	Late ALP Esztár Observed %	Late ALP Esztár Ppd
I1	1.23	- 0.12				
I2	2.47		1.49			
I3	8.65		10.45	- 1.40		
I4	22.22		41.79	-10.30	36.11	- 4.62
II1	22.22	-10.38	35.82	-23.98	19.45	- 7.61
II2	11.11		8.96		33.33	-21.73
II3	25.93	- 6.91	1.49		11.11	
II4	6.17					
Total Ppd		17.41		35.68		33.96

Vegetation zones	Late ALP Szilmeg Observed %	Late ALP Szilmeg Ppd	Late ALP Bükk Observed %	Late ALP Bükk Ppd	Late ALP Szakálhát Observed %	Late ALP Szakálhát Ppd
I2			6.41	- 1.79		
I3			15.38	- 6.33		
I4	32.14	- 0.65	42.31	-10.82		
II1	14.28	- 2.44	26.92	-15.08	1.22	
II2	21.44	- 9.84	8.98		1.22	
II3	17.86				79.27	-60.25
II4	14.28	- 3.01			18.29	- 7.02
Total Ppd		15.94		34.02		67.27

Percentage figures for the regional groups of the Late ALP are shown in Histograms k, l, m, n, o and p (Fig. 2). The percentages of settlements in the dry forest-steppe are: 32 in the Late AVK; 1 in the Tiszadob; 11 in the Esztár; 32 in the Szilmeg; 0 in the Bükk; and 98 in the Szakálhát Group. The percentages of sites in the transitional forest-steppe zone are: 33 in the Late AVK; 45 in the Tiszadob; 53 in the Esztár; 36 in the Szilmeg; 36 in the Bükk; and 2 in the Szakálhát Group. The rest of the settlements are in the forest zone: 35 in the Late AVK; 54 in the Tiszadob; 36 in the Esztár; 32 in the Szilmeg; 64 in the Bükk; and 0% in the Szakálhát.

The results show clearly contrasting patterns for the Körös culture and the TLP. While Körös sites are entirely absent from the forest zone, the majority of TLP settlements are located there. The ALP seems to occupy a transitional position in this respect. Its settlements are found in both zones. In contrast with the Körös culture, fewer ALP settlements are in the dry forest-steppe, and the majority of the sites are found along the steppe margins. In the Late phase of the ALP strong regional differentiations are clearly noticeable. Again, it is the Szakálhát group which is most similar in its distribution to the Körös culture. In the TLP, settlements, with the exception of a very few sites in the Late phase, are absent from the dry forest-steppe zone. In Late TLP, there seems to have been a slight shift from forest to steppe margin occupation, although the number of sites in the forest zone is still relatively high. This shift to drier areas was already noticed when settlement distributions according to climate regions were examined.

Chi-square tests performed for the three cultures suggest that it is extremely unlikely that the observed distributions could have arisen by chance; the probability is less than 0.05%. Percentage point differences between the expected and observed patterns are shown in Table 5. Some of the values, especially those for the ALP, are relatively low. This, on the one hand, is due to the existence of strong chronological and regional differences within the ALP; on the other, the low values are a reflection of the transitional character of this culture. In the case of the ALP, the zonal vegetation approach is limited in its usefulness to explain settlement distribution.

While the distribution of settlements according to vegetation zones is informative as to the degree of forest occupation, zonal vegetation maps are too generalised to show finer relationships between settlements and local vegetation. As has been pointed out by plant ecologists, vegetation development even in the natural or wild state, that is without any human interference, does not always reach the climatic climax stage. Where natural long-term arresting factors, such as drainage impedence or extreme steepness of land, are present the prisere (series of plant communities which, in turn, occupy an originally fresh untenanted surface) is prevented from proceeding to the climatic climax. An almost permanently arrested community is referred to as a subclimax (Eyre, 1963). The Natural Vegetation Map of Hungary includes both climax and subclimax communities. It attempts to show the 'natural' plant cover of the country as it would have looked before the clearance of forests, tilling of the soil, control of the major rivers, and artificial drainage and reclamation of marshlands (National Atlas of Hungary, 1967).

The limitation of any vegetation map is that they show only the final communities in the succession. In reality fresh mineral surfaces are constantly being exposed within the general areas of all types of climatic climax (or subclimax) vegetation so that many types of priseral vegetation, at various stages of development, are to be found (Eyre, 1963, 16). In Hungary, for instance, large areas were flooded annually before the regulation of the rivers, and thus new surfaces were created either by deposition or erosion. To supplement the information of the map the plant associations of Hungary showing the successive developmental stages from the simpler to the more advanced are described in some detail.

The Plant Associations of Hungary

1. Aquatic and floodplain plant associations: Before the regulation of the rivers and drainage of the marshes,there were extensive areas of floodplain and swamp vegetation,especially in the Great and Little Plains along the Danube and the Tisza and their tributaries. Marshes were characteristically found around Lakes Balaton, Fertö and Velence.

Aquatic-marshy associations: The first community which invades the near shore areas of standing or sluggish waters is composed of floating water-grasses. The dominant species are duckweeds (*Lemna minor* and *L. gibba*), small water fern and floating moss (*Salvinia natans*). The next association is that of large water grasses (*Potamogeton*), which are rooted at the bottom. The more frequent species are the water buttercup (*Ranunculus aquatilis*), the water caltrop (*Trapa natans*) and the white water lily (*Nympha alba*). As silting continues,reeds (*Scirpo-Phragmitetum*) appear, then tall sedges (*Magnocaricion*), and finally the plant associations of swampy meadow (*Agrostion albae*), shrub-willows and the forest associations of the floodplain.

Floodplain forest associations: The banks, oxbows and shoal islands of the rivers are lined with willow and poplar groves (*Saliceto-Populetum*). Willow and poplar groves are characteristically found on the lower floodplains. The foliage stratum of this soft wood association is composed of the white willow (*Salix alba*), the crack willow (*Salix fragilis*) and the black poplar (*Populus nigra*). In the shrub stratum,hawthorn species dominate, and the most frequent species in the herbaceous layer is the dewberry (*Rubus caesius*).

On the higher floodplain, which is inundated only by the larger floods, the succession reaches the stage of elm-ash-oak grove forest (*Querco-Ulmetum hungaricum*). The foliage stratum of this hardwood association is dominated by the pedunculate oak (*Quercus robur*), the field elm (*Ulmus campestris*) and the sharp-pointed ash (*Fraxinus oxycarpa*). In the shrub stratum,the characteristic plants are hawthorns, dogwoods and occasionally the tartar maple (*Acer tataricum*). The wild grape, *Vitis silvestris*, is also found at this level. The more common species of the herbaceous layer are *Brachypodium sylvaticum*, *Urtica dioica*, *Corydalis cava*, *Geranium robertianum*, *Euphorbia stricta*, *Rumex sanguineus*, *Festuca gigantea*, etc. They are characteristically found on moist but well-drained soils which are high in nitrogen (Zólyomi and Simon, 1969). A schematic representation summarising a typical river-side succession is shown in Fig. 3.

Grass-moor associations: These associations are found away from the floodplains in stagnant waters and undrained depressions. Local shallows are inhabited by tussock and tall sedge associations. The dominant species is clump sedge (*Carex elata*). Tussocks are succeeded by wet grass-moor associations formed by fenland sedge (*Carex davalliana*) and large rush (*Juncus subnodulosus*). This is a peat forming association. The end association is alder-moor forest. The characteristic species are the grey willow (*Salix cinerea*), the resinous alder (*Alnus glutinosa*) and the sharp-pointed ash (*Fraxinus oxycarpa*). These fen woodlands or shrublands are commonly known as 'carr'.

2. Plant associations of the forest-steppe:

Associations on alkaline soils: A large part of present-day alkaline soils belonged before the regulations to the floodplain area. Areas of alkaline soils nonetheless existed before the draining of the floodplains and marshes. The plant associations of the alkali flats vary according to the water supply and the sodium content of the soil.

Where soil alkalinity is low alkali pastures are formed. The dominant species is sodic fescue (*Festuca pseudovina*). As alkalinity increases,the sodic fescue is replaced by sagebrush (*Artemisia maritima*), which in turn is replaced by salt-meadow grass (*Puccinellia limosa*). The poorest association is that of *Camphorosma ovata*.

The final association on alkali soils is the tartar maple-oak forest (*Galatello-Quercetum roboris*). It is usually found on slightly higher elevations than the floodplain, in a position which is transitional between the floodplain and the higher loess plateaus. On the one hand, it adjoins the elm-ash-oak grove of the floodplain, on the other the alkali meadow steppe. In this transitional belt, soil alkalinity is not very high and is usually absent from the surface horizon (solonetz profile). It has been suggested that as a result of the downcutting of the Tisza in the Subboreal, the elm-ash-oak grove was in many places replaced by alkaline forests (Zólyomi and Simon, 1969).

Associations on sand: Wind blown sand dunes occupy large areas between the Danube and the Tisza, and in the Nyirség. The first plants to invade unanchored sand are brome grass (*Bromus tectorum*) and woodrye (*Secale silvestre*). Beside the grasses lower order plants, mosses and lichen also grow on sand. Partially stabilized slopes are occupied by Hungarian fescue (*Festuca vaginata*) together with *Ephedra distachya*. The next phase in the succession is the closed meadow steppe formation. The dominant species here is the striated fescue (*Festuca sulcata*). In depressions, poplar and juniper appear as pioneers in the steppe. The final association consists of two types of oak woods. The first is the lowland oak (*Festuca-Quercetum roboris*) which is drier and more open. The second is the oak with the lily of the valley (*Convallario-Quercetum roboris*). It occurs where the ground water table is higher, and is often found in the proximity of the elm-ash-oak grove forest.

Associations on loess: As the loess areas of Hungary have been cultivated for a long time, the reconstruction of the 'natural' vegetation is somewhat

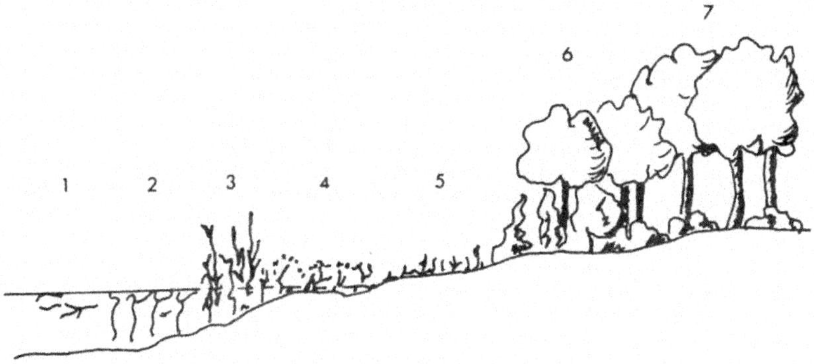

Fig. 3. Cross section of a typical river-side succession (from Pécsi and Jakucs, 1971).

1, floating reed grass; 2, fixed reed grass; 3, reeds; 4, sedge; 5, swamps; 6, grove forest with Salix and Populus; 7, grove forest with Ulmus, Fraxinus and Quercus.

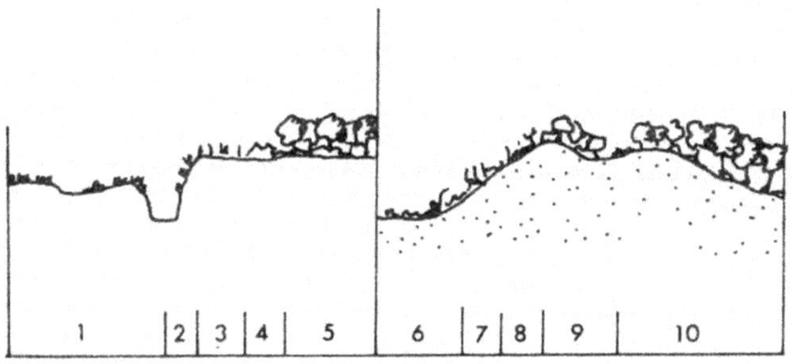

Fig. 4. Cross section of steppe vegetation on loess and sand (from Pécsi and Jakucs, 1971).

1, vegetation on alkali soil; 2, Agropyro-Kochietum on loess; 3, loess steppe; 4, shrub with Amygdalus nana; 5, oak woods with tartar maple; 6, Salicetum rosmarinifoliae; 7, sandy steppe with Festuca vaginata; 8, sandy steppe with Festuca sulcata; 9, sparse oak woods on sandy soil; 10, canopied oak forests on sandy soil.

tentative. The initial plant association was the closed grass steppe. The grasses included the striated fescue (*Festuca sulcata*) and the curly needle grass (*Stipa capillata*). At the contact of loess and forest steppes,the shrub association of dwarf almonds (*Amygdaletum nanae*) was very widespread. The final asosication was the loess steppe forest of tartar maple and oak (*Aceri tartarico-Quercetum*). The dominant trees were the pedunculate (*Quercus robur*) and hairy oak (*Q. pubescens*), occasionally the turkey oak (*Q. cerris*), the field elm (*Ulmus campestris*) and the tartar maple (*Acer Tartaricum*). The typical associations on loess and sand are summarised in Fig. 4.

3. Associations of the deciduous forest zone

Karst forest association: A good example of arrested prisere development is found on the Mezozoic dolomite blocks of the Hungarian Mountains. On the steeper slopes only rock grass associations can survive. Where the soil cover is thicker,the rock grass is replaced by steppe meadows. The final association is a karst shrub forest (*Cotino-Quercetum* or *Ceraso-Quercetum*). This is not a true forest; it is rather park-like. The dominant species are the hairy oak (*Quercus pubescens*), the flowery ash (*Fraxinus ornus*), the Hungarian fustic (*Cotinus coggygria*), the *Prunus mahaleb* and *Cornus mas*.

Oak forest associations: The most widespread forest association in Hungary is the oak forest. Oak forests are typically found below 600 m in elevation, usually on the sunnier southern slopes of the hills and mountains. Limestone slopes are occupied by calciphile oak forests (*Orno-Quercetum* or *Corno-Quercetum*). This closed forest often accompanies the karst shrub-forest association. The foliage level is composed of hairy oak (*Q. pubescens*), sessile oak (*Q. petraea*) and turkey oak (*Q. cerris*).

The *Quercetum petrae-cerris* association is found over the most varied bedrocks in deep clayey or sandy soil. Areally, this is the most widespread forest association in Hungary. The calcifuge oak association (*Genisto tinctoriae-Quercetum*) is mainly found on volcanic slopes. In the foliage stratum, the birch accompanies the sessile oak.

In cooler, moister areas, usually between the oak and beech forest zone, hornbeam-oak forest are the typical associations. As the oak does not cast a really heavy shade,oak forests usually have rich and varied shrub and herbaceous layers.

Beech forest associations: Beech forests are relatively limited in extent in Hungary. The beech likes a cool climate and relatively abundant precipitation. Beech forests are generally found between 600-900 m in Hungary. The shrub stratum is generally absent,and the carpet level is sparse because of the heavy shade of the beech.

Beech forests with woodruff (*Asperula*) are found on cooler northern slopes or higher limestone plateaus. On southern rocky slopes, the characteristic type is the beech forest with *Melica*.

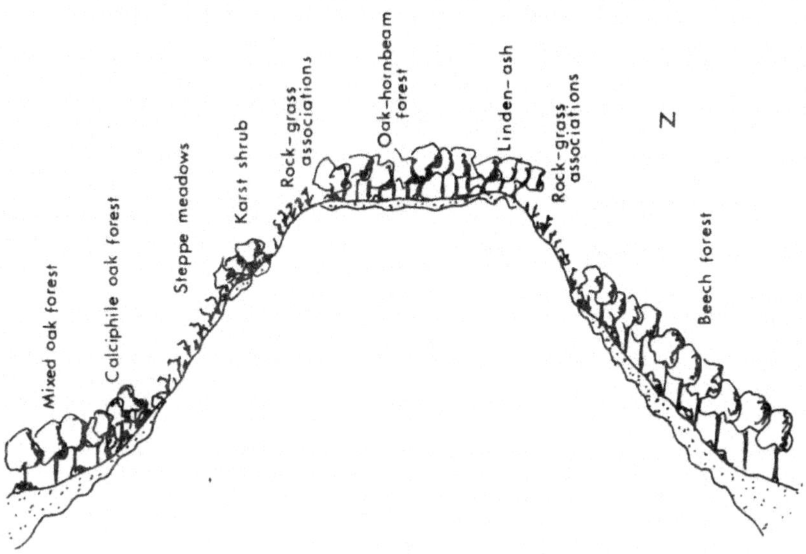

Fig. 5. Vegetation zonation on a typical low mountain in Hungary (from Pécsi and Sárfalvi, 1964).

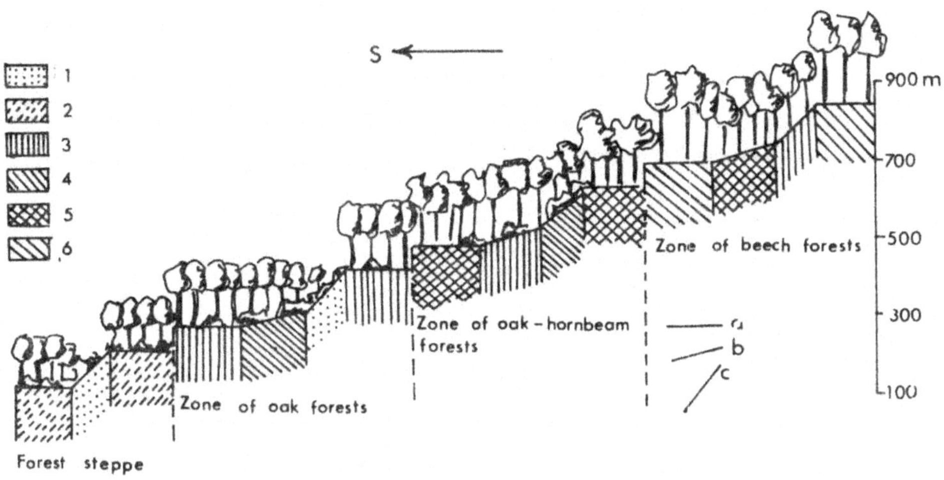

Fig. 6. Forest zones on southern slopes of different gradients (from Pécsi and Jakucs, 1971).

1, steppe and shrub forest; 2, loess oak forest; 3, turkey oak-sessile oak forest; 4, calci- and thermophilous oak forest; 5, hornbeam-oak forest; 6, beech forest. Slope gradient: a) 0 to 5°; b) 10 to 20°; c) 20 to 35°.

MAP 5. Settlement Distributions and Natural Vegetation

Key

1. Oak woods on sandy ground, sandy steppe (Festuco-, Convallario- Quercetum roboris)
2. Oak woods with tartar maple on loess (Aceri tatarico-Quercetum pubescenti- roboris)
3. Loess steppe with shrub, with spots of oak woods on loess and on alkali soils
4. Complex of maple-oak woods in the lowlands (Aceri campestri-Quercetum petraeae-roboris)
5. Tartar maple-oak woods on alkali soil (Galatello-Quercetum roboris)
6. Solonetz vegetation (formerly and partly a floodplain vegetation)
7. Solonchak vegetation
8. Riparian forests in the floodplains (mainly Fraxino pannonicae-Quercetum roboris); swamps and bogs
9. Hornbeam-oak forests (Querco robori-Carpinetum)
10. Swamps and swamp forests
11. Karst forest of Illyrian affinity with hairy oak (Orno-Quercetum pubescentis), karst shrub (Cotino-Quercetum pubescentis) and rocklawn association
12. Hairy oak forest of continental character (Corno-Quercetum pubescenti petraeae), shrub forests (Cerasο-Quercetum pubescentis), rock lawns, slopes with steppe-type grass vegetation
13. Acidiphilous oak forests (Quercetum petraeae cerris, Potentillo-Quercetum, Tilio argenteae-Quercetum)
14. Montane hornbeam-oak forest (Querco petraeae-Carpinetum)
15. Submontane beech forest (Melico-Fagetum)
16. Montane beech forest (mainly Aconito-Fagetum)
17. Beech forest of Illyrian affinity (Helleboro-, Vicio oroboidi-Fagetum)
18. Hornbeam-oak forest of Illyrian affinity (mainly Helleboro-Carpinetum)
19. Scotch pine-oak forest (Pino-Quercetum)
20. Scotch pine forest with whortleberry (Myrtillo-Pinetum)
21. Silver fir-spruce forest (Abieti-Piceetum)

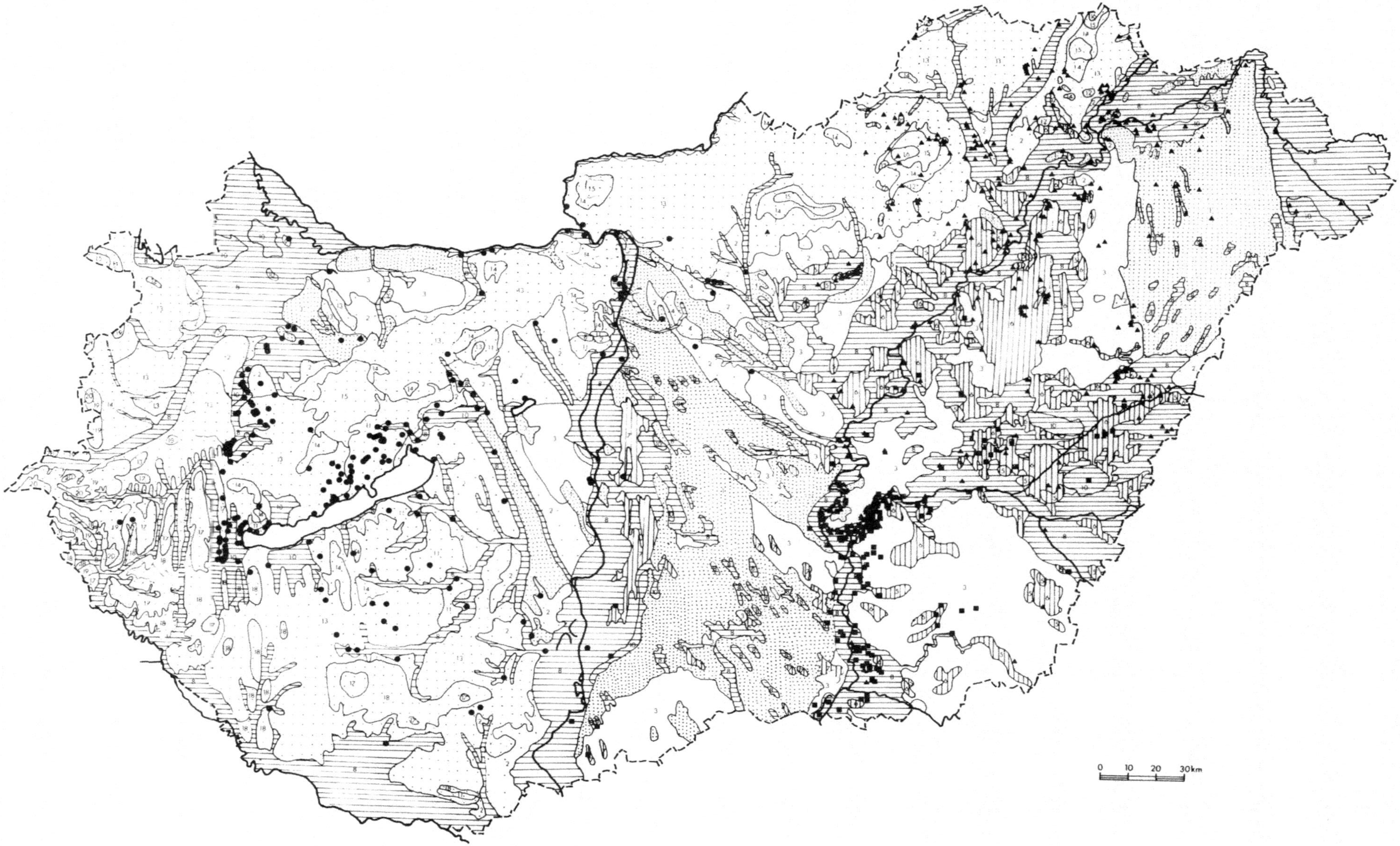
0 10 20 30km

Fig. 7 Histograms showing the distribution of settlements in relation to plant associations.

1, oak woods on sand, sandy steppe (Festuco-, Convallario-Quercetum roboris)
2, oak woods with tartar maple on loess (Aceri tatarico-Quercetum pubescenti-roboris)
3, loess steppe with shrub
4, lowland maple-oak woods (Aceri campestri- Quercetum petraae-roboris)
5, tartar-maple oak woods on alkali soil (Galatello- Quercetum roboris)
6, solonetz vegetation (formerly and partly a floodplain vegetation)
7, solonchak vegetation
8, floodplain forest (Fraxino- Quercetum roboris); swamps
9, hornbeam-oak forest (Querco robori-Carpinetum)
10, swamps and swamp forests
11, karst forest of Illyrian affinity with hairy oak (Orno-Quercetum pubescentis), karst shrub (Cotino-Quercetum pubescentis), rockgrass
12, hairy oak forest of continental character (Corno-Quercetum pubescenti petraeae), shrub forests
13, acidiphilious oak forests (Quercetum petraeae cerris)
14, mountain hornbeam-oak forest (Querco petraeae-Carpinetum)
15, low-mountain beech-forest (Melico-Fagetum)
16, mountain beech forest (Aconito-ragetum)
17, beech forest of Illyrian affinity (Helleboro-, Vicio oroboidi-Fagetum)
18, hornbeam-oak forest of Illyrian affinity (Helleboro-Carpinetum)
19, Scotch pine-oak forest (Pino-Quercetum)

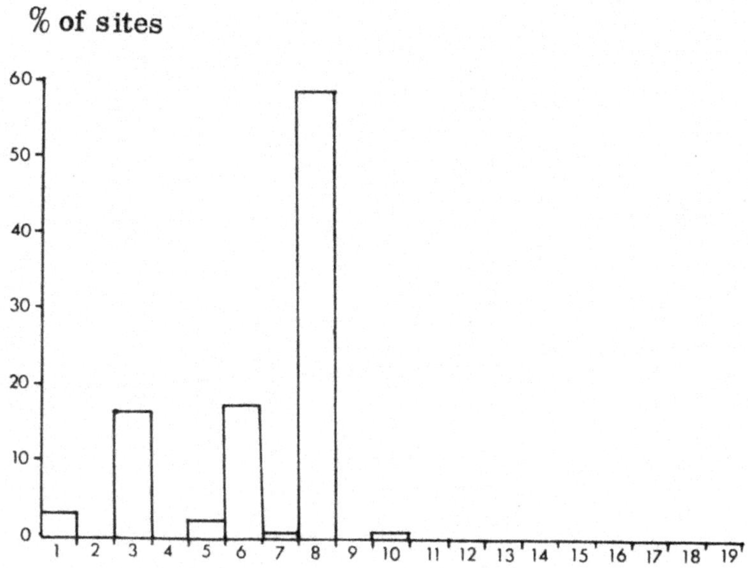

a) Körös culture

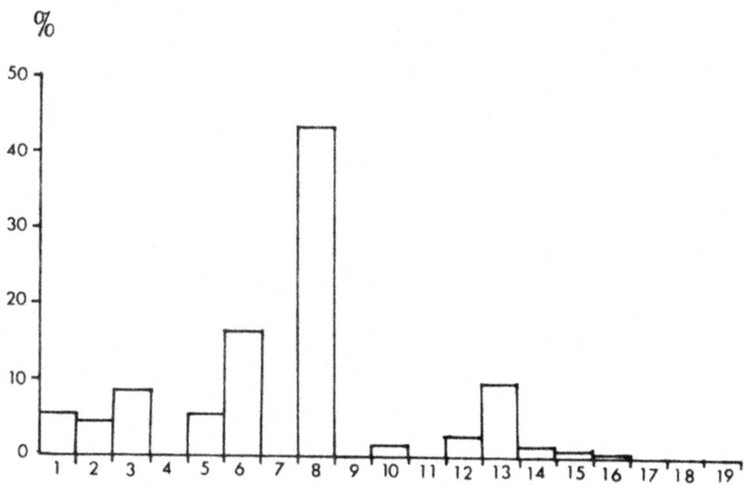

b) Alföld Linear Pottery culture

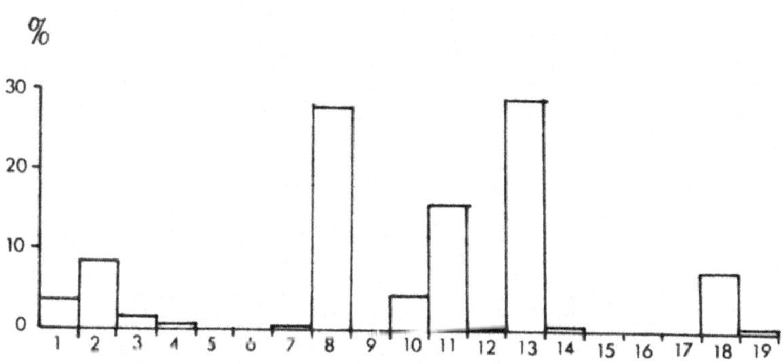

c) Transdanubian Linear Pottery culture

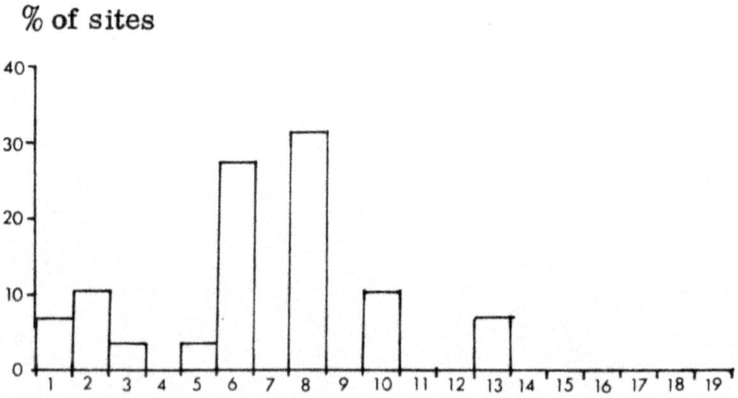

d) Early ALP

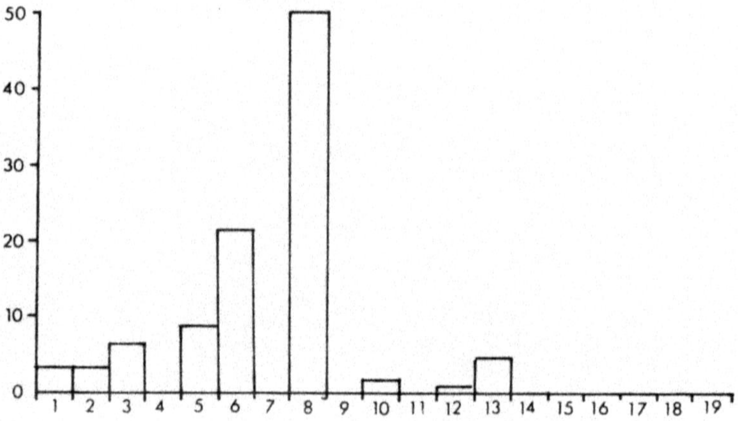

e) Middle ALP

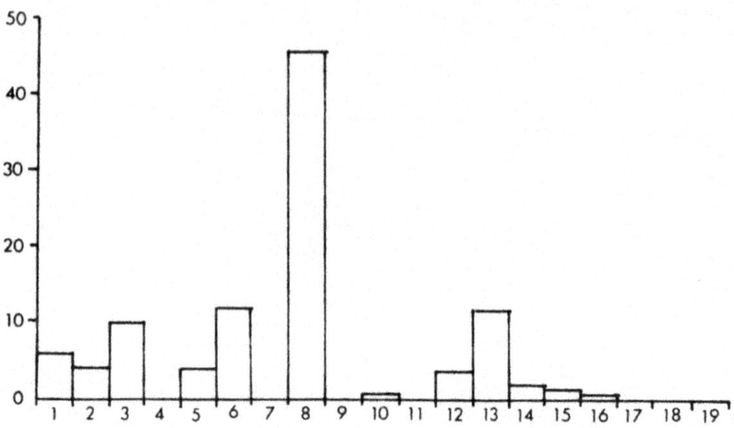

f) Late ALP

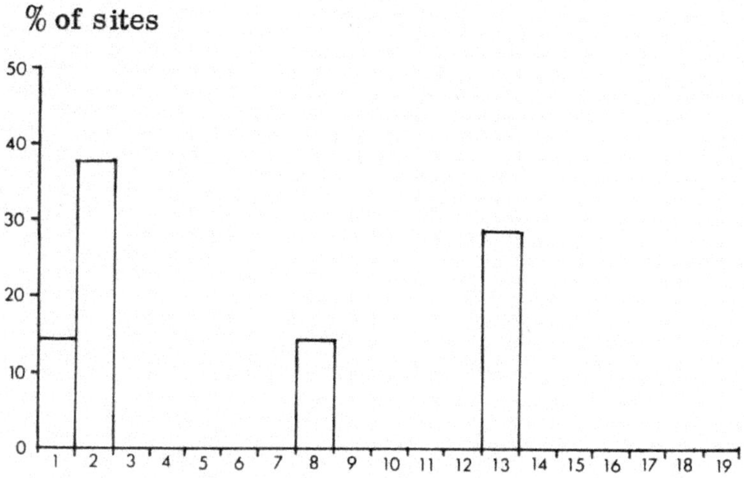

h) Early TLP

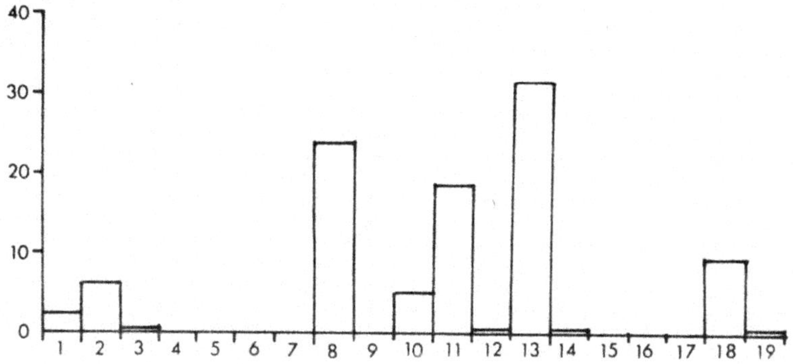

i) Middle TLP

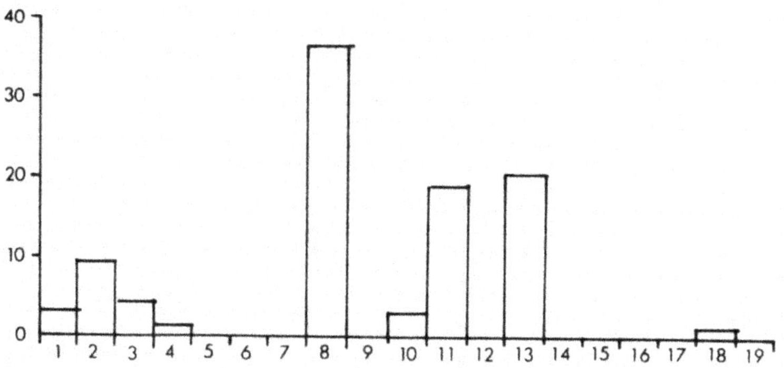

j) Late TLP

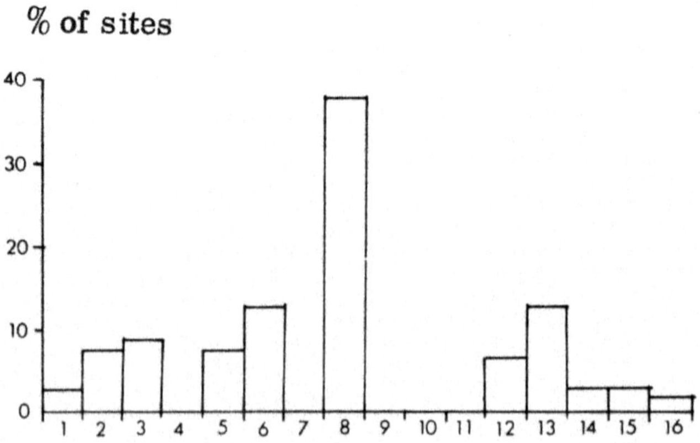

k) Late AVK

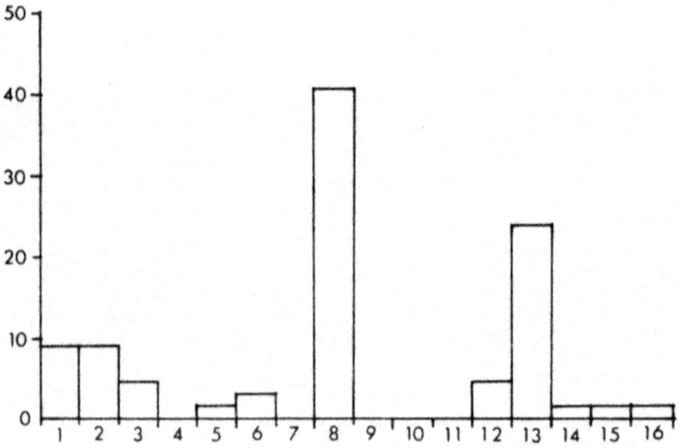

l) Tiszadob

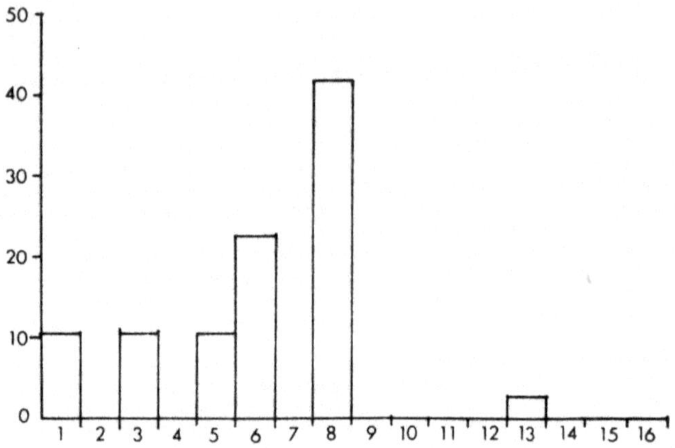

m) Esztàr

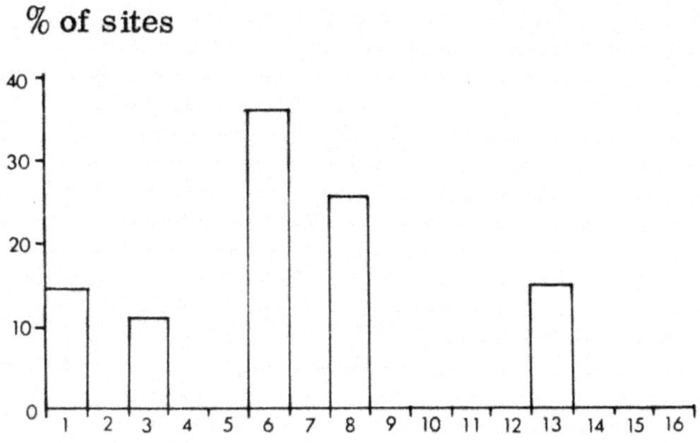

n) Szilmeg

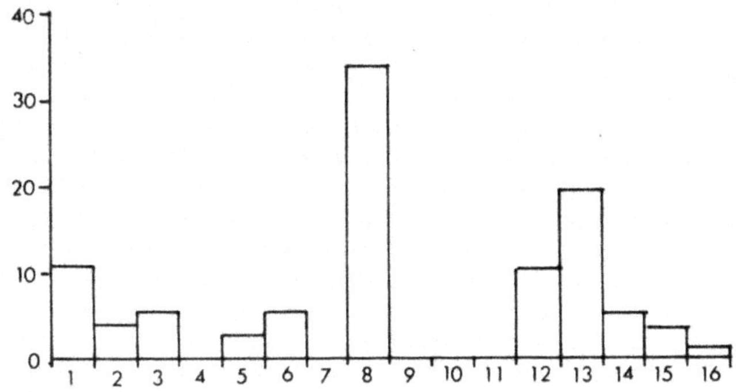

o) Bükk

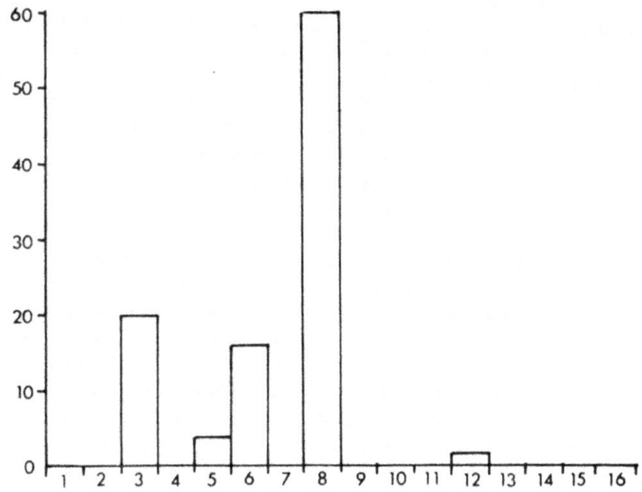

p) Szakàlhàt

Table 6. Percentage point difference between expected and observed distributions of settlements in relation to vegetation

Vegetation	Körös culture Expected %	Körös culture Observed %	Körös culture Ppd	ALP Observed %	ALP Ppd	TLP Observed %	TLP Ppd
1	14.51	3.06		5.45		3.77	
2	8.04			4.16		8.49	- 0.45
3	11.95	16.79	- 4.84	8.84		1.42	
4	0.22					0.47	- 0.25
5	1.41	2.29	- 0.88	5.45	- 4.04		
6	5.30	17.56	-12.26	16.10	-10.80		
7	1.22	0.76				0.47	
8	23.34	58.78	-35.44	43.63	-20.29	27.83	- 4.49
9	0.05						
10	2.42	0.76		1.30		4.25	- 1.83
11	1.85					15.57	-13.72
12	0.46			2.86	- 2.40	0.47	- 0.01
13	19.50			9.87		28.77	- 9.27
14	3.21			1.30		0.47	
15	0.98			0.78			
16	0.11			0.26	- 0.15		
17	1.41						
18	2.88					7.55	- 4.67
19	0.84					0.47	
20	0.27						
21	0.03						
Total Ppd			53.42		37.68		34.69

Ppd = Percentage point difference

Vegetation	ALP Early Observed %	ALP Early Ppd	ALP Middle Observed %	ALP Middle Ppd	ALP Late Observed %	ALP Late Ppd
1	6.90		3.18		5.95	
2	10.34	- 2.30	3.17		3.97	
3	3.45		6.35		9.92	
5	3.45	- 2.04	8.73	- 7.32	3.97	- 2.56
6	27.59	-22.29	21.43	-16.13	11.90	- 6.60
8	31.03	- 7.69	50.	-26.66	45.24	-21.90
10	10.34	- 7.92	1.59		0.40	
12			0.79	- 0.33	3.57	- 3.11
13	6.90		4.76		11.51	
14					1.98	
15					1.19	- 0.21
16					0.40	- 0.29
Total Ppd		42.24		50.44		34.67

Table 6 (contd.)

	TLP					
	Early		Middle		Late	
Vegetation	Observed %	Ppd	Observed %	Ppd	Observed %	Ppd
1	14.28		2.51		3.17	
2	42.86	-34.82	6.29		9.52	- 1.48
3			0.63		4.76	
4					1.59	- 1.37
8	14.28		23.90	- 0.56	36.51	-13.17
10			5.03	- 2.61	3.17	- 0.75
11			18.87	-17.02	19.05	-17.20
12			0.63	- 0.17		
13	28.58	- 9.08	31.45	-11.95	20.64	- 1.14
14			0.63			
18			9.43	- 6.55	1.59	
19			0.63			
Total Ppd		43.90		38.86		35.11

	Late ALP					
	Late AVK		Tiszadob		Esztár	
	Observed %	Ppd	Observed %	Ppd	Observed %	Ppd
1	2.47		8.96		11.11	
2	7.41		8.95	- 0.91		
3	8.64		4.48		11.11	
5	7.41	- 6.	1.49	- 0.08	11.11	- 9.70
6	12.35	- 7.05	2.99		22.22	-16.92
8	37.04	-13.70	40.30	-16.96	41.67	-18.33
12	6.17	- 5.71	4.48	- 4.02		
13	12.34		23.88	- 4.38	2.78	
14	2.47		1.49			
15	2.47	- 1.49	1.49	- 0.51		
16	1.23	- 1.12	1.49	- 1.38		
Total Ppd		35.07		28.24		44.95

	Late ALP					
	Szilmeg		Bükk		Szakálhát	
	Observed %	Ppd	Observed %	Ppd	Observed %	Ppd
1	14.29		10.25			
2			3.85			
3	10.71		5.13		19.51	- 7.56
5			2.56	- 1.15	3.66	- 2.25
6	35.71	-30.41	5.13		15.85	-15.55
8	25.	- 1.66	33.33	- 9.99	59.76	-36.42
12			10.26	- 9.80		
13	14.29		19.23		1.22	
14			5.13	- 1.92		
15			3.85	- 2.87		
16			1.28	- 1.17		
Total Ppd		32.07		26.90		61.78

The characteristic associations on the southern and northern slopes of a typical low mountain are shown in Fig. 5, and the forest zones of southern slopes of different gradients are summarised in Fig. 6.

Pine woods in Hungary are presumably not indigenous, but were introduced by man after afforestation. The vegetation of Hungary is very rich in species. It includes, due to the country's position and climate, Oceanic, Mediterranean and Continental elements (Pécsi and Jackucs, 1971). The richness of the flora is also due to the fact that Hungary was never entirely covered by ice and thus many Tertiary and Pleistocene species have survived.

Settlement Distributions and 'Natural' Vegetation

The distribution of settlements according to vegetation types is shown on Map 5. Percentage values for the Körös ALP and TLP settlements associated with each type are given in Histograms a, b, and c (Fig. 7). Histograms d, e, f, h, i and j (Fig. 7) show the percentage distributions in the Early, Middle and Late phasesof the ALP and TLP; Histograms k, l, m, n, o, p (Fig. 7) give the percentage figures for the regional groups of the Late ALP.

The most noticeable feature of settlement distribution according to vegetation type is the importance of the azonal gallery forest association for the location of the sites, especially in the Körös and ALP cultures. The majority of Körös sites (59%) and also a large percentage of the ALP sites (44%) are in the gallery forest. It has been pointed out previously that some of the areas which are now characterised by solonetz vegetation would have, prior to the regulation of the rivers, carried gallery forests or swamp vegetation. It is also possible that the tartar-maple woods on alkali soils were gallery forest areas in the Atlantic. If the solonetz vegetation and tartar-maple woods are also counted as gallery forests, the percentage figures are 79 for the Körös culture and 65 for the ALP. As some alkali vegetation would have been present even before the regulation of the rivers, the real figures for gallery forest occupation are probably between the two extreme values.

To some extent, the attraction of the gallery forest for settlement was the attraction of the river or watercourse to which this azonal environment was due. But beyond that the gallery forest area itself had a number of advantages for settlement. In the dry forest steppe, where moisture is limiting, the swamp edges and gallery forests were areas with adequate soil moisture. The importance of these areas for agriculture will be discussed later in greater detail. In addition, swamps, backswamps and gallery forests carried rich wild faunas, birds and mammals. Thirdly, although the forest steppe was not devoid of trees, the best supply of building material and fuel was the gallery forest. There is good archaeological evidence to indicate that the trees of the gallery forests were utilised by the Körös and ALP people. Macroscopic analysis of fossil wood remains from the Körös sites of Kopáncs and Kotacpart Vata tanya and the Linear Pottery and Tisza site of Kisköre showed species typical of the gallery forests: Ulmus and Quercus at the Körös sites and Quercus, Ulmus and Fraxinus at Kisköre (Hollendoner, 1933-34; 1935).

In the TLP, where the majority of the sites is in the forest zone, the importance of the gallery forest is much less (28%). In the Late TLP, however, where a slight shift toward steppe margin occupation has been observed, there is also a slight increase in the occupation of the gallery forest (37%).

A relatively large percentage of settlements in the Körös culture (17%) and in some of the regional groups of the Late ALP are found on the loess steppe with shrub vegetation. It should be pointed out, however, that none of the sites is in the middle of the steppe; all the settlements are located at the edge of the steppe, next to the floodplain. A more accurate term for the description of their location would be the ecotone zone between the steppe and gallery forest.

In the forest zone, the majority of settlements, both in the ALP and TLP, is found in various oak wood associations. In the ALP, only 1% of the sites are in the beech zone, and in the TLP sites are absent from this zone altogether. In the Bükk culture, there is a slight increase in the occupation of beech forests, but the percentage figure is still very small (5%).

The highest percentage of settlements in the oak zone is associated with the acidiphilous oak forest (Quercetum petrae and cerris): 10% in the ALP 29% in the TLP. The percentage of settlements in calciphile oak wood associations is relatively high in the TLP (16%) and in the Bükk culture (10%). The occupation of the more mesophile hornbeam-oak forest is more marked in the TLP (8%) and is very slight in the ALP (1%). One interesting point about the late groups of the ALP (with the exception of Late AVK and Szakálhát groups) is the increase of the number of settlements in oakwoods on sand. This together with the extension of settlement into the beech forest zone suggests an increased interest in more marginal areas. A slight increase in the number of settlements in oakwoods on sand is also noticeable in the Early phase of the ALP as compared to the Körös culture.

The preference of the early agriculturalists for the oak wood associations in the forest zone is hardly surprising. Oak forests occupy the sunnier southern slopes and are found at lower elevations. They have a much richer undergrowth layer and consequently a better supply of wild food for man and fodder for animals. They are also found on better soils than the beechwoods in Hungary. It is possible that in the selection of the oak forest areas considerations such as soil type or subsoil properties were the deciding factors, but the nature of the vegetation cover would nonetheless have been an important factor in influencing both farming and grazing practices. The likely effect of the local vegetation on the economy will be discussed later in greater detail.

Chi-square tests for the distributions of the three cultures suggest that it is extremely unlikely that the observed patterns would be due to chance alone; the probability of chance distribution in either culture is less than 0.05%. Percentage point differences between the expected and observed figures are shown in Table 6. The values for the Körös and TLP cultures are similar to those obtained by the zonal approach. In the case of the ALP, the figures are generally higher.

Vegetation in the Atlantic Phase

On the natural vegetation map of Hungary an attempt is made to show the plant cover of the country as it would have looked had not the clearance of the forests, tilling of the soil, control of the rivers and reclamation of marshlands occurred. The assumption is that the climatic controls are those of today. In the following part, the vegetation in the Atlantic phase is described, and an attempt is made to assess how far the real distribution would have differed from the theoretical one that is shown on the map.

The reconstruction of the Atlantic vegetation is based chiefly on the evidence of pollen diagrams and the analysis of macroscopic plant remains. The geographical locations where pollen cores were taken is shown in Fig. 8. Pollen diagrams in Hungary are as yet few. Furthermore, while certain regions such as the Balaton and Nyirség are well represented, in others pollen diagrams are lacking altogether. This is partly due to the bad preservation of pollen in certain sediments such as sand and loess. Another disadvantage is that none of the pollen sequences has as yet been radiocarbon dated. This introduces some element of uncertainty in the cross dating of the sequences. But in spite of the shortcomings, the evidence is adequate to allow a tentative reconstruction of the Atlantic vegetation cover.

The pollen diagrams from Lake Balaton show that the region around the lake had a relatively open mixed oak forest cover (Zólyomi, 1953). The dominant species were the Quercus, Ulmus, Tilia and Fraxinus (altogether 40%). The percentage of Corylus in the total AP count is 15%. In the second half of the Atlantic the Ulmus, Tilia and Fraxinus decrease, while the percentage of Fagus and Carpinus pollen begins to rise. Karst steppe and karst forest associations (Brometum and Querceto-Cotinetum) were also present.

In the Nyirség region, according to the pollen diagrams, the dominant association was also the mixed oak forest with Fraxinus, Tilia and Ulmus. In moister areas, alder was predominant. Corylus values fluctuate and show several minor peaks during the phase. In the second half of the Atlantic, the percentage of the mixed oak forest increases to 53 from 22-41 (Csinády, 1959), but among the individual trees Tilia and Fraxinus show a decrease. In some of the diagrams, Fagus and Carpinus are already present in the first phase (Csinády, 1969-60; Borsy and Borsy, 1955), but in general they appear only in the second phase. In all the pollen profiles, NAP values (as a percentage of the total AP) are relatively high: 38% at Csaroda (Csinády, 1959); 58-66% at Kokad in the second half of the Atlantic (Csinády, 1959-60); 34-40 at Döge; 20-65% at Kiskálló (Borsy and Borsy, 1955); and 14-54 at Bátorliget (Csinády, 1954). Some of the high values for NAP are due to the increase in aquatic plants as bog formation accelerated.

At Ócsa and Dunakeszi, the Atlantic phase is again characterised by the dominance of the mixed oak forest (Járai-Komlódi, 1966). The values for alder are similarly high (48% at Dunakeszi). Corylus shows two maxima during the phase. Ilex, Hedera and Vitis appear for the first time in Hungary. NAP values fluctuate between 7 and 17%.

The pollen diagram from Alpár includes only the last phase of the Atlantic (Járai-Komlódi, 1966). Mixed forests with Quercus and Ulmus are dominant. Alder is again strongly represented. Vitis, Hedera and Viscum are also present. The AP count is 74%.

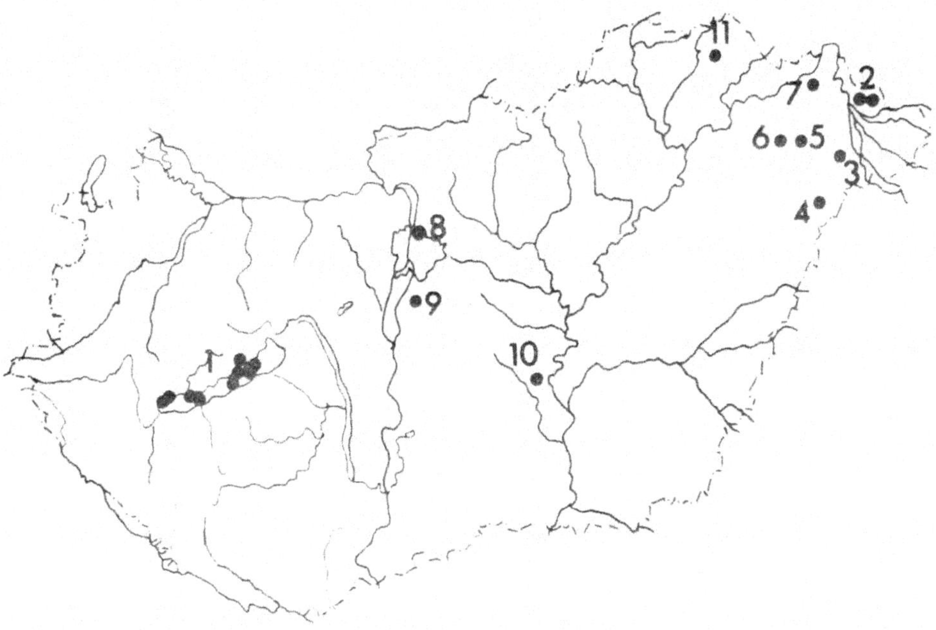

Fig. 8. Locations of pollen profiles in Hungary. (Only profiles which contain the Atlantic sequence are included.)

1, Lake Balaton; 2, Csaroda and Csaroda-Nyirestó; 3, Nyirbátor-Bátorliget; 4, Kokad; 5, Kiskálló; 6, Nyiregyháza-Ujfehértó; 7, Dőge; 8, Dunakeszi; 9, Ócsa; 10, Alpár; 11, Tokártető.

The swamp at Tokártető is in the beech zone today. The pollen diagram shows that the dominant forest here in the Atlantic was the mixed oak with Quercus, Corylus and Ulmus. (Járai-Komlódi, 1971.) Beech was also present. In spite of the closed nature of the forest (AP: 60-90%),pollens of heliophilous species typical of a steppe vegetation (Artemisia, Helianthemum, Ephedra,etc.) are also present. According to Járai-Komlódi,they represent rock grass associations which must have been present on the slopes by the Hernád and the steppes of the Hegyalja. Tokártető was a small Sphagnum bog during the Atlantic.

Fossil wood remains from Neolithic sites in the Alföld have been identified as Quercus (mostly Q. robur but also some Q. cerris), Ulmus, Fraxinus, Acer, Salix, Betula and Corylus (Zólyomi, 1958). At Békásmegyer, near Budapest, Quercus has been found in the Neolithic layer (Hollendoner, 1935), and in the Hillebrand cave in the Bükk mountains, Quercus, Fraxinus and Populus have been identified also from the Neolithic deposit (Stieber, 1956).

On the basis of pollen diagrams and macroscopic identifications, (Zólyomi, (1958) has attempted to reconstruct the Atlantic vegetation of the Northern Hungarian Mountains and the Alföld. The reconstruction is shown in Fig. 9. The cross section gives the characteristic associations of several microenvironments: northern and southern slopes, foothills, loess steppe, sanddune and floodplain. For comparison, the cross sections of the preceding (Boreal) and subsequent (Subboreal) periods are also included.

The reconstruction of the vegetation of the forest zone is relatively secure. The vegetation cover here was not greatly different from that shown on the Natural Vegetation Map. Mixed oak forests with Hedera and Ilex in places predominated. Karst steppe and forest associations were characteristic of rocky slopes. The main difference seems to have been the restricted role that Carpinus and Fagus played in these associations. Mixed oak forests probably extended beyond their present day limit of 500-600 m , and the occurrence of the hornbeam and beech was rather sporadic.

There is little doubt that floodplain and alder carr associations were also widespread during the Atlantic. On the higher floodplain, in the oak-elm-ash forest grove Vitis silvestris was already present. Although there is no direct pollen evidence for the presence of alkali vegetation in the Atlantic, their existence, though limited, is assumed on other grounds (Somogyi, 1965).

On sand, the succession, at least in places, reached the stage of Convallario-Quercetum roboris association. Hazel was abundant in the shrub layer (Járai-Komlódi, 1966). The appearance of sand dunes in the Alföld is dated to the Preboreal (Zólyomi, 1958).

The ecological status of the loess areas of the Alföld has been the subject of long debate among ecologists (Soó, 1931; 1959). According to the present consensus of opinion, the Alföld was a forest-steppe interspersed with treeless meadows on the loess uplands during the Atlantic phase (Járai-Komlódi, 1966). Pollen diagrams show the existence of very extensive steppe vegetation in the preceding Boreal phase; the percentage of nonarboreal pollen reaches 90% in the Balaton diagrams (Zólyomi, 1953). While the rate of colonisation by trees was no doubt greatly accelerated by the moister climate of the Atlantic, pollen

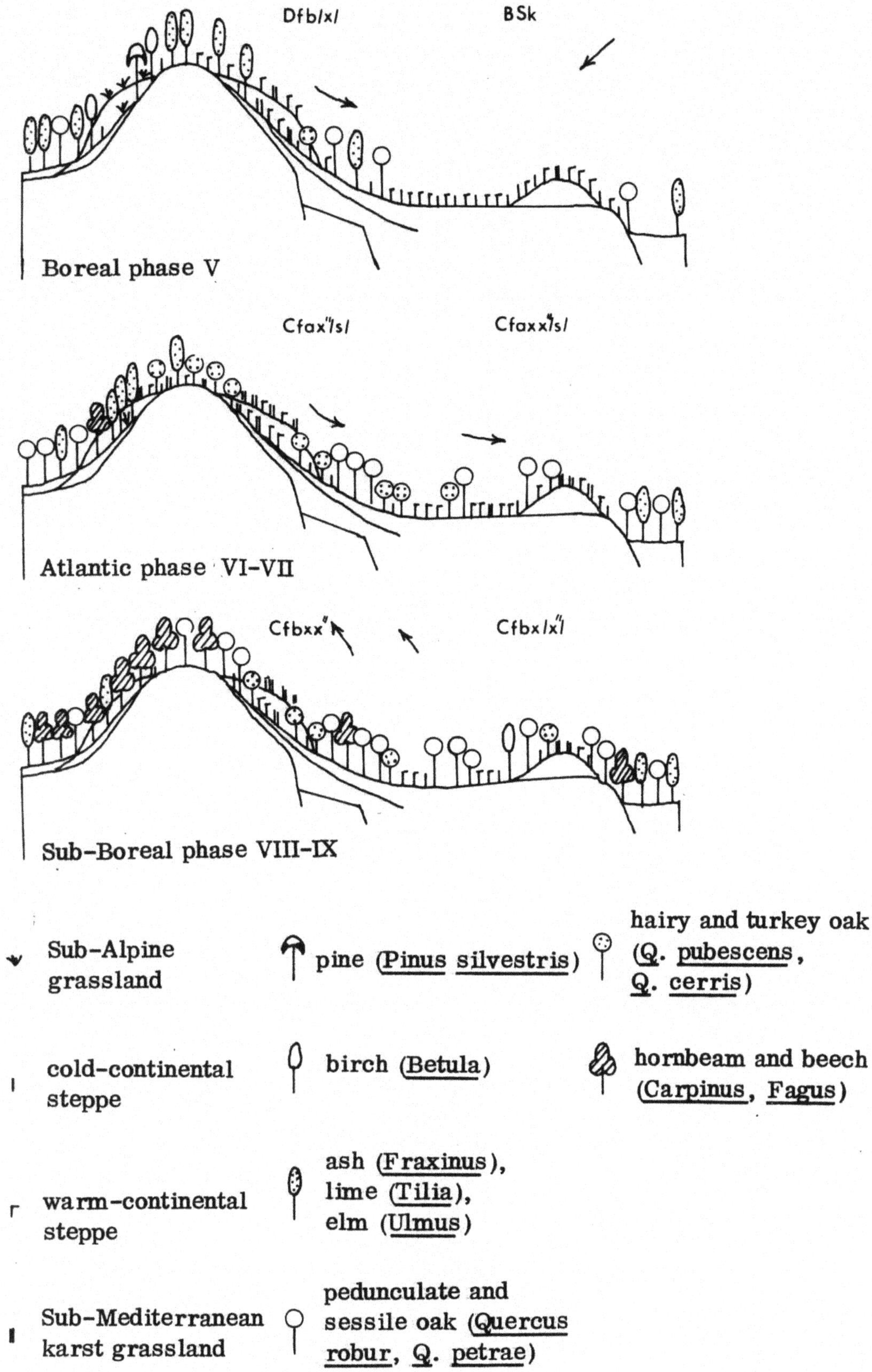

Fig. 9. Vegetation and climate zones in the Northern Mountains and the Alföld (after Zólyomi, 1958).

diagrams do not suggest the existence of dense forests in the Alföld at this time. Pollen profiles from the margins of the Alföld usually include pollens of typical steppe vegetation, such as Artemisia, Gramineae, Chenopodiaceae. Evidence from other disciplines also suggests the existence of a relatively open forest steppe vegetation, at least in parts of the Alföld. The faunal evidence has already been mentioned. Soil studies, discussed later, would tend to support the same view. In contrast to Poland, where according to Kruk (1971) the loess areas carried closed forest vegetation, or the western part of Hungary where the loess was also well wooded, the Alföld is best characterised as a mosaic of forest-steppe and floodplain and marsh forest vegetation. An additional factor in slowing down the reforestation of the Alföld was the appearance of agricultural man.

While man's influence at this time was still relatively small, the impact of his activities may have been more pronounced in an ecologically transitional area, such as the forest-steppe of the Alföld. Ložek (1967) attributes a similar role to man in the driest low-lying areas of Czechoslovakia. He suggests that the chernozem steppe in these areas cannot be regarded as a result of deforestation caused by human interference; man's influence was more in the nature of protecting already existing forest-steppe enclaves from the invasion of the forest. Most of the pollen diagrams in Hungary show some evidence of the presence of man. Pollens of cultivated cereals appear for the first time, though they are still small in quantity. Weeds of cultivation are also present in some of the diagrams: Plantago lancelota, Pl. major-media, Polygonum avic tip., and Rumex at Alpár; Polygonum avic tip. at Ócsa and Dunakeszi (Jarai-Komlódi, 1966). It is possible that the increase of Betula and later of pine pollen together with the increase of Pteridiophyta species in some of the Nyirség diagrams (Csinády, 1954, 1959) may have been at least in part due to the activities of man. The Atlantic was also the period when the immigration of Mediterranean species reached its maximum in Hungary (Zólyomi, 1958). Although climatic factors were mainly responsible, it is not unlikely that man's agency probably partly unintentionally, also played a part in the dispersal of certain wild species beyond the Mediterranean.

C. HYDROGRAPHY

Hungary belongs to the drainage area of the Danube. The Danube is the largest river in Central Europe with a catchment area extending over 9 countries. The second largest river in Hungary is the Tisza, which having run most of its course in this country, joins the Danube at Titel, Yugoslavia.

The Danube and its tributaries

The Danube enters Hungary in the northwest and maintains an easterly course till it reaches Vác. Around Vác, the river turns to the south and flows in a southerly direction till it leaves the country below Mohács. When the Danube enters Hungary, it has no well defined valley; it flows in a braided pattern across its alluvial fan, the Little Plain. From Komárom the Danube has a single channel and a well defined terraced valley. Between Budapest and Mohács the valley becomes much wider; in the east it is bordered by an earlier alluvial fan of the Danube, the Danube-Tisza Interfluve, and in the west toward the Mezőföld plain by fault-controlled and steeply undercut bluffs

(Pécsi, 1959). The Danube can be described as a graded river, although there are some sections where aggradation and again others where degradation is dominant.

The discharge of the Danube in Hungary is determined by the flow of the Upper Danube, which is largely dependent on the discharge of its right bank tributaries from the Alps. The annual precipitation over the catchment areas of these Alpine rivers is very high: 1200-2000 mm (Pécsi and Sárfalvi, 1964). The largest volume of water is contributed by the Inn. Depending on the fluctuation of Alpine precipitation, the time and intensity of snowmelt in the mountains, the Danube in Hungary has two regular floods each year: one in early spring, the other in early summer.

The early spring flood of April or May is due to the melting of the snow in the upper reaches of the river and its tributaries. The vehemence of the earlier flood is usually less than that of the later summer flood, even though the water recedes only slowly as there is constant recharge from the melting snow. The ice on the Hungarian section of the Danube is usually broken up by the spring flood.

The early summer or 'green flood' is due to the late spring or early summer precipitation maximum in the catchment area of the Upper Danube. In addition, the melting of the snow in the higher Alpine glaciers also starts at this time and contributes to some extent to the discharge of the rivers. The Danube reaches its highest water level in Hungary in June. The maximum discharge at Budapest is 10 000 cubic m/sec and the minimum is 600 (Pécsi and Sárfalvi, 1964). The difference between the highest and lowest levels can be as much as 8 m. The Danube is lowest towards the end of the summer and in early autumn. Winter levels are also relatively low.

The right bank tributaries of the Danube in Hungary are the Lajta, Rába, Répce and the Marcal in the Little Plain and the Sárviz in the Mezőföld. Before the Sárviz joins the Danube it takes up the waters of the Sió and the Kapos. All these rivers are characterised by relatively small discharge. The only right bank tributary of appreciable discharge is the Dráva, which joins the Danube in Yugoslavia. The Dráva has 3 floods; in addition to the early spring and early summer floods, which are also characteristic of the other tributaries, there is also a third flood in early autumn due to a second maximum of precipitation over its catchment area. The left bank tributaries are less significant; their discharge is smaller. The largest river is the Ipoly, which empties into the Danube at the border with Czechoslovakia.

The Tisza and its tributaries

The Tisza is also a river at grade, that is erosion and deposition are in balance. The course of the Tisza through the Great Hungarian Plain is a meandering one; the channel is accompanied by cutoffs, cutoff lakes or oxbows and silted up fens.

The section of the river between the Romanian border and the mouth of the Bodrog is usually referred to as the Upper Tisza Riverine. It consists of two marginal subsided basins: the Szatmár-Bereg plain and the Bodrogköz-Rétköz. From Tokaj to the confluence with the Körös the floodplains of the

Tisza and its tributaries form the Middle Tisza Riverine. The Lower Tisza Riverine is the lowest lying surface in the Great Plain; the elevation of this area does not exceed 78-80 mm a.s.l. (Pécsi, Somogyi and Jakucs, 1971). The valley of the lower Tisza is a trough-like depression, which is bordered by sharp bluffs. In the south, a narrow flood-free terrace strip follows the western margin of the floodplain.

The Tisza also has two floods: one in early spring, the other early summer. The early summer flood, as on the Danube, is higher and more vehement. The early spring flood is due to the melting of the snow in the Carpathian and Transilvanian mountains. The flooding of the tributaries, which occurs at the same time, contributes to the increase in the discharge of the Tisza. The early spring floods of the Danube and the Tisza also frequently coincide. As the Tisza cannot discharge into the Danube, disastrous floods occur further upstream. The ice on the Tisza usually disappears before the early flood arrives.

The early summer flood is due to the late Spring-early summer precipitation maximum over the catchment area of the Tisza. The floods of the Tisza and its tributaries generally coincide but precede the early summer flood on the Danube. The highest stage is usually reached by the end of May. The maximum discharge at Szolnok is 3 800 cubic m/sec and the minimum is 72 (Pécsi and Somogyi, 1971). It is characteristic of the extreme regime of the Tisza that the fluctuation in water level is 3 to 4 meters greater than it is on the Danube. East of Tokaj the Tisza usually has a third flood due to a second maximum in precipitation in October.

The right bank tributaries of the Tisza in Hungary are the Bodrog, Sajó, Eger and Zagyva. The most important is the Bodrog, which is the union of five smaller rivers. The left bank tributaries are the Kraszna, Szamos, Körös and Maros. The Maros is the largest, but only a short section of it flows through Hungary. The Körös, after which the Körös culture was named, carries the waters of five smaller rivers into the Tisza. Like the Tisza, it is a graded river; it has a meandering course and has often changed its bed.

The drainage network of Hungary on the whole is not very dense. Certain areas, such as the sand dune areas of the Great Plain, have insufficient drainage. In Transdanubia and in the Hungarian mountains, there is a fair number of smaller water courses, but many of these are intermittent and dry out during periods of drought.

Lakes and marshes

The largest lake in Hungary is Lake Balaton. The graben depression the lake occupies is relatively young; it was formed at the end of the Pleistocene (Pécsi and Somogyi, 1971). The northern shore is narrow and rocky, while the shore line in the south is formed by a wide flat sandy beach.

The water surface of the lake is approximately 600 sq km. Its greatest length is 77 km; its greatest width 14 km. The average depth is 3-4 metres. The fluctuation of the water level is no more than 2 meters; it is highest in April or May and lowest in October or November.

The Balaton is too small to affect the climate of more than its immediate environment. The influence is most noticeable on the southern slopes of the north shore, where the reflection of sunshine from the water increases insolation.

The third largest lake is Lake Velence; the second largest is Lake Fertő, but most of the lake lies outside Hungary. The surface of Lake Velence is only 26 sq km. Before drainage, however, the lake also included extensive marshes to the southwest.

Beside the three larger lakes there are several smaller ones in Hungary. According to the cadaster of the Scientific Research Institute for Water Conservation, the country has about 1200 standing waters having a surface greater than 0.5 hectare (Pécsi and Somogyi, 1971).

Before the regulation of the rivers and the drainage works of the last century, there were several extensive marshes in poorly drained areas. The largest ones were the Ecsed Swamp and the Bodrogköz in the Upper Tisza Riverine; the Small—and Great Sárrét between the Berettyó and the three Körös rivers; the Kisbalaton by the river Zala; and the Hanság in the Little Alföld. These marshes once supported rich wildlife. They also served as resting places for migrating birds since Hungary lies along the route of the annual spring and autumn migration of northern and east European birds (Pécsi and Somogyi, 1971).

Ground Water

Of the subterranean waters only ground water is mentioned here. As a rule, the depth of the ground water table in loess areas corresponds to the thickness of the loess layer. In the Great Plain or in the Mezőföld, it can be 6 to 12 meters or even more. In areas of aeolian sand and in river valleys, the ground watertable is usually found between 2 to 5 metres. In lower lying floodplains it is usually even nearer the surface. The fluctuation in the level of the ground water table is usually small under the loess uplands (1-1.5 m) and large near the rivers (6-7 m) (Rónai, 1961).

In the Great Plain, where the ground water moves mostly in fine grained calcareous sediments, the water is hard and usually calcareous. In consequence, it is hardly suitable for drinking. Because of the large amount of soluble salts in it, the ground water has been an important agent in the formation of alkali soils over large areas in the Alföld (Szabolcs, 1961).

The regulation of the rivers and draining of the marshes in the last century had the effect of lowering the ground water table over large areas. According to soil investigations in the Alföld, the drop in the water table was 0.4-1.5 m in the Sárrét, 0.5-0.8 m in the Hortobágy, 0.1-1 m around Szeged and 0.3-0.5 m in the Kiskunság (Somogyi, 1965). The magnitude of the change is not very great, but it had important consequences for soil formation.

Settlement Distributions in Relation to Rivers and Lakes

Water has traditionally been an important factor in settlement location; empirical data collected by Chisholm (1968) suggest that nearness to water

MAP 6. Settlement Distribution and Hydrography Prior to River Control

Key

Area inundated for most of the year

Area inundated at times of flood

0 10 20 30 km

usually overrides other considerations in the placement of settlements. Rivers and lakes in addition to supplying water for man and beast are also important as food resources and means of transport and communication. Before the invention of the wheel the attraction of the river route was no doubt even greater.

To estimate the importance of rivers and lakes in the location of the Körös and Linear Pottery sites, the average distance to the nearest river or lake was calculated for each culture, and the results were compared with the mean distance calculated for Hungary as a whole. The mean for each culture was calculated by measuring the distance of each site to the nearest river or lake and then by totalling and dividing the results by the number of sites. To calculate the mean value for Hungary, the whole country was divided into 5 km squares, and the distance from the centre of each square was then measured to the nearest river or lake. The results thus obtained were totalled and divided by the number of squares. If nearness to rivers was an important factor in the selection of sites, the means for the three cultures would be expected to differ significantly from the mean obtained for the whole of Hungary. To test whether the difference between the means could be due to chance t-tests were performed.

The mean values for the three cultures are as follows: 0.83 km in the Körös culture; 0.97 km in the ALP; and 1.22 in the TLP. The mean value for Hungary is 2.32 km. t-tests show that the difference between the means of the three cultures and the mean for Hungary is highly significant; it is extremely unlikely that the location of the settlements near rivers and lakes could be the result of chance distribution.

Although the mean values calculated for the three cultures show that the settlements were located near rivers, the figures are relatively high and give an overestimation of the real distances involved. This is due to the fact that, for practical purposes, the measurements were taken on a relatively small scale (1:1 000 000) map, from which the smaller rivers had naturally been omitted. If a smaller area is examined in greater detail, the correspondence between site location and river courses is much closer.

Kalicz (1957) in his survey of the Tiszazug region has noted that the Körös sites are, without exception, found immediately by the rivers. The sites follow the course of the river along a narrow strip; the width of the settlements is usually very small in relation to their length. In contrast, Linear Pottery sites in the same area are found both by and a little further away from the rivers, even if the distance (150 m) is still very small. A different emphasis on resources or a difference in the function of the settlements could account for the slight shift in site location, although the degree of difference is small and should not perhaps be overemphasised.

The regulation works of the last century have shortened and changed many of the river courses; the length of the Tisza, for example, was cut by 38% (Lászlóffy, 1969), and the other rivers, especially on the Great and Little plains, suffered a similar fate. Because of the drastic nature of the changes, the distances measured were those to the pre-regulation river courses (Map 6). As seen on the map, the settlements follow very closely the old pre-regulation courses. This would suggest that the modern drainage pattern, small differences notwithstanding, was already in existence in the Atlantic period.

Settlement Distributions and Floodplains

It has been mentioned that most rivers in Hungary have two regular floods each year. Before the regulations, large portions of the country were inundated by the floods: one million hectares along the Danube and over two million along the Tisza were temporarily or permanently under water (Pécsi and Somogyi, 1971). Two-thirds of the Great and Little Plains were occupied by floodplains. The distribution of the settlements in relation to the floodplain is shown on Map 6.

To provide a relative measure of the importance that floodplains played in settlement location, the percentage of inundated land was calculated around each settlement within a 5 km square area. The figures obtained for the individual sites were totalled and divided by the number of settlements. According to the calculations,70% of the land in the immediate vicinity of the Körös sites was inundated every year; the average figure in the ALP is 56%; and in the TLP 26%.

The high proportion of inundated land around the ALP and especially the Körös sites suggests a large degree of reliance on the resources of the floodplain, among which water is perhaps the most valuable. It has been mentioned in the section on climate that a large percentage of the ALP and all Körös sites are located in areas which are climatically deficient in moisture. While rainfall may have been limited in these areas, the immediate environment of the sites seems to have been fairly rich in surplus water during most of the year. Inadequate rainfall could be supplemented in the floodplain either by utilising the receding floodwaters of early spring and early summer, or by using at high water stage the elevated ground water table near the rivers for crop production. In the TLP, where moisture is far less limiting, the importance of the floodplain is also less; on the average only a small percent of the land in the immediate vicinity of the sites is occupied by flood plain environment. The comparable figure in the ALP, as we have seen, is twice as much, and in the Körös culture three times as much.

While floodwaters can be beneficial, they can also be destructive to settlement and crops. To understand more fully the potentialities and limitations of the floodplain environment for settlement it is necessary first to give a brief description of the characteristic features of the floodplain.

Floodplain morphology

Floodplains are constructions of alluvial rivers. Alluvial rivers are those with beds and banks composed of solids deposited by the river, in contrast to those with channels across solid bedrock. Alluvial rivers often change their courses and, at times, abandon previously used courses for completely new ones. At flood stage, the alluvium is deposited most rapidly along the immediate sides of channels, where the deposits form ridges, called natural levees. 'The two most significant parts of the floodplain are the natural levees, along most alluviating channels, and low areas between them, basins.' (Russell, 1967).

Natural levees: Levees vary in dimension. The elevation of their crest is determined by floodstage in the adjacent channel. The banks facing the river are relatively steep, but the backslope of wide natural levees is relatively gentle and forms a gradually flattening surface toward the adjacent basin.

As levees rise above the rest of the floodplain, they are inundated only by the highest floods. When the floodwaters begin to recede, the levees are immediately left dry, while the low lying basins remain under water. In Hungary some of the drainless basins carry water all through the year (Map 6).

When the river is in flood the coarser and heavier load is deposited first. Hence the levee deposits are invariably coarser than those of the basins. The nature of the deposit depends on the load of the individual river. The deposits of the Tisza consist of middle to fine grained sands and silts. The Danube carries a lot of gravel when it enters Hungary; most of the heavy load is deposited in the Little Plain. Because levees are coarser in texture they have good subsurface drainage.

The ground water level in the levees is low during low water stage and rises when the level of the river is high (Buringh, 1957). There is always a natural subsoil drainage of ground water in the levees in the direction of the basin at high water—and toward the river at low water stage. As a consequence, even in saline or alkaline areas where the groundwater is generally rich in dissolved salts, the ground water under the levees is usually fresh and free of salts. Due to the presence of the ground water table within reach of trees and woody plants, levees can support timber growth even in relatively dry areas (Butzer, 1965).

Basins or backswamps: In comparison with the levees, the basins are low and wet. If covered by trees, they are called swamps and, if by grass, marshes. This distinction is usually overlooked, however, and as a rule the relatively flat areas between natural levee systems are collectively regarded as backswamps (Russell, 1967). The basins accumulate large quantities of organic sediment because of the vigorous marsh or swamp growth. As a result, they usually become beds of peat or peaty clay that are waterlogged in many cases. These layers somtimes become impregnated by solids derived from salts dissolved in both surface and ground water. The most frequent tree in the backswamp is the alder.

Settlement Location and Floodplain Morphology

It is quite clear from the short description of levee and basin characteristics that levees and related features,such as point bars and alluvial tongues, offer numerous advantages for settlement in the floodplain. Because they are higher than the surrounding plain, levees are relatively safe from the danger of floods. Their backslopes provide favourable conditions for surface drainage; when floods overtop them, the water drains off quickly. Levees also offer some protection against crop-killing frosts,which frequently settle into low-lying areas during the winter (MacAdams 1965). Composed of alluvial deposits, levee soils are usually fertile and because of their lighter textures are also easier to work. They usually have good subsurface drainage and at the same time are also well supplied with moisture from the fluctuating ground water table. In contrast to the depressions, the ground water of the levees is salt free. Finally, levees have the important advantage of being nearest to the river.

In view of all the advantages that levees offer for settlement, it is hardly surprising that most of the Körös and also many of the ALP sites, at least in the Lower and Middle Tisza riverines, are in fact found on natural levees. Nandris (1970) has observed the association of the Körös sites with 'islands of low water table'. In most cases, these islands are former levees, and the low water below them is usually a reflection of surface topography. In the surrounding basins,the ground water table is as a rule nearer the surface.

Describing the living conditions of ancient times in the Lower Mesopotamian Plain, Buringh (1957) has observed that there are certain differences between arid and humid regions in the location of settlements in river plains. 'The ancient settlements in river plains of the arid regions are all formed in or at the border of the river basins. In contradiction to this, those in the humid regions are formed on the river levees. In humid areas under natural conditions,the basins always are too wet and therefore unsuitable for living and cultivation. In arid regions, however, the levees are unsuitable for cultivation as they are too dry to grow crops under natural conditions. Ancient villages are found mainly in the basins. This is also the case in the Adana Delta of Turkey'. Buringh's example of a humid region is Holland, where such a pattern of levee crest occupation has frequently been observed in the river clay areas. Settlements in the arid regions are still often found on the levees (MacAdams, 1965; Butzer, 1965), but their position in relation to the basins is different; they are usually nearer the basins.

Conditions in the Great Hungarian Plain are similar to those found in Holland; the basins are usually too wet, and settlements prefer the levee tops. There is, however, an important difference between the two countries; in Holland floods occur mostly in winter (Edelman, 1950), while in Hungary the time of high water stage, varying somewhat according to the individual river, is from early spring to early or mid-summer. While the high water stage lasts, the elevated ground water table of the levees, either directly or through the capillary zone,is within reach of the root zones of most plants. Thus, quite independently of rainfall a steady supply of moisture is assured for the crops for a limited period of time. The influence of the timing and duration of the floods on crop policy and the effect of the relative shortage of suitable dry land on agriculture in the floodplain will be discussed later.

The settlements of the Körös culture are found, as mentioned before, in a long narrow strip along the crest of the levees. The ALP sites, though also located on levees, show greater variation in the positioning of the settlements. In some instances, sites occur on very slight elevations, in others, much nearer to the flooded depression (Kalicz, 1957; Korek, 1961). This greater diversity in the location of the ALP sites may have been due to a greater differentiation in the function of the settlements although the evidence is as yet too slight to warrant very definite conclusions.

Levees, although naturally well protected from prolonged inundations, are, nonetheless, vulnerable to floods. Flood danger was an ever present threat to the Körös and ALP sites in the Alföld. In some cases, the flood may have been disastrous enough to lead to the abandonment of the settlement, at least temporarily. This seems to have been the case at the Körös sites of Hódmezővásárhely-Kopáncs-Zsoldos tanya, Dévaványa-Réheli Gát and at the later Neolithic site of Hódmezővásárhely-Kökénydomb. At these, and probably at many others, the site was abandoned after a layer of inundation silt had been deposited. When the flood was less vehement, however, the levee position offered some protection. Ethnographical evidence from the Alföld suggests that the usual wattle and daub house construction was in a sense a form of adaptation to withstand smaller floods without having to abandon the site (Székely, 1961). The flood washed away the daub but usually left the wooden framework more or less intact. After the flood had receded, the frame was replastered.

It has been mentioned, that alluvial rivers often change their beds. Before the regulations, the rivers in the plains of Hungary were no exceptions. The Danube, Tisza and their tributaries often changed their courses across the Alföld. Csalog (1941) mentioned that at Medina the site, when excavated, was below the present level of the ground water table. As an explanation, climatic change has been postulated. While climate may have been partly responsible, a shift in the course of the Sárviz can just as well have brought about the change. A change in the course of the river would naturally be of importance for settlements. Not only would the distance to the river change, but the drainage conditions of the area would also be altered. Papp (1960) cites the example of the Small and Great Sárréts, where changes in the drainage conditions of the Berettyó and Körös rivers led to significant changes in settlement patterns within relatively recent historical times. Because of alterations in the course of rivers and consequent changes in the growth of swamps, several villages had to be abandoned and many had to move their location. Between Keményháza and Szerep, for example, the course of the river shifted approximately 3 km within 4 centuries. As a result, Keményháza was abandoned and never rebuilt, while Szerep was moved to a new location.

Similar changes must have occurred during the Atlantic period. There is little doubt that the relatively great mobility of the Körös and ALP sites can be at least partially attributed to the frequent alterations in the courses of the rivers in the Alföld. Other factors, however, were probably involved. The degree of permanency of occupation differs in the Körös and Linear Pottery cultures, which as the sites of the two cultures are often found in similar areas cannot have been entirely due to environmental conditions.

The Evolution of the Drainage Network in Hungary

The previous analysis of settlement distributions was made on the assumption that the pre-regulation river network did not significantly differ from that of the Atlantic period. The close correspondence between pre-regulation river courses and the location of the Körös and Linear Pottery sites suggests that the assumption is to a large extent justified. Nonetheless, according to geomorphological research certain changes in river morphology did occur after the Atlantic period. In the following part, the evolution of the Hungarian drainage network is described briefly, and an attempt is made to assess how far the changes postulated for the post-Atlantic period would have altered Atlantic hydrography.

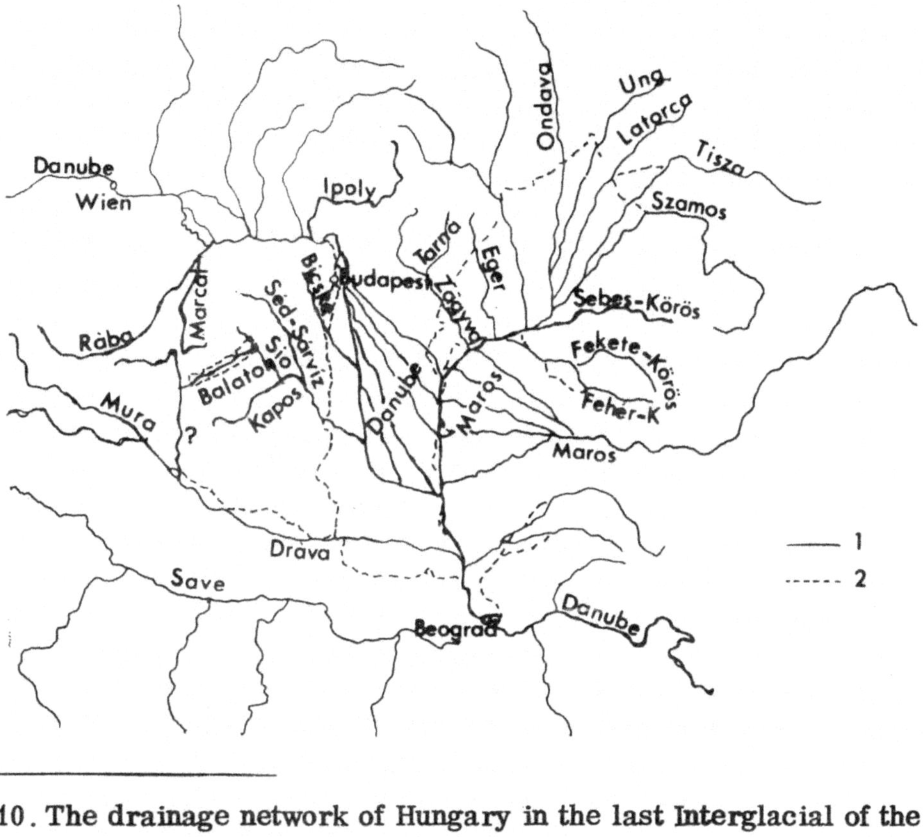

Fig. 10. The drainage network of Hungary in the last Interglacial of the Pleistocene (from Somogyi, 1967)

1, Pleistocene drainage network; 2, Present-day drainage network.

The present drainage network began to evolve at the end of the Pliocene, when the Pannonian-Pontian sea, which had previously occupied most of Hungary finally disappeared (Somogyi, 1967). The ancient Danube filled up the basin of the retreating Pannonian Sea in the area of the Little Plain. The river then turned south and, proceeding along the valley of the present-day Dráva, ended in a system of inland lakes in the southern part of the Great Plain. Later, due to the tectonic uplift of the central portion of Transdanubia, the course of the Danube was diverted toward the Visegrád gorge. Emerging from the gorge, the Danube built up a vast alluvial fan at the edge of the subsiding basin of the plains. The reconstructed drainage network of Hungary in the last Interglacial of the Pleistocene is shown in Fig. 10.

As seen on the map, neither the Danube nor the Tisza followed its present course in the Great Plain at the time. The Danube, broken up into several branches, flowed southeast toward Szeged; the Tisza occupied the valley of the Ér-Berettyó-Körös rivers and flowed toward the present Lower Tisza valley. The Danube occupied its present north-south valley as the result of a small subsistence at the border of Transdanubia at the end of the Pleistocene. The Tisza valley between Tokaj and Szolnok came into existence due to marginal subsidences in the area of the Upper Tisza Riverine in the early postglacial period. The drainage network of Hungary acquired its present shape essentially by the end of the Preboreal period.

During the Boreal the rivers were aggrading (Somogyi, 1967). As a result a large amount of alluvium was deposited along the river channels. In the Great Plain most of the alluvium consisted of fine grained loessic silt eroded from uncovered Pleistocene surfaces.

Climatic change and denser vegetation cover in the Atlantic resulted in the reduction of aggrading activity. During this period the rivers formed huge meander loops in the plains. Both pollen and archaeological evidence dates these meanders to the Atlantic phase. Many of the Körös and Linear Pottery sites are found along them.

The climate of the following Subboreal phase in Hungary was, in contrast to Scandinavia, wetter and cooler (Zólyomi, 1958). Greater rainfall and decreased evaporation led to an increase in stream discharge and finally to valley incision (Somogyi, 1962). The extent of the down-cutting was relatively small, 2-4 m at the most (Somogyi, 1965). The formation of the new terrace nonetheless,resulted in important changes in floodplain morphology. The Atlantic floodplain was now 3-4 m above the level of the rivers. Although still reached by the highest floods, it was no longer under the direct influence of the rivers. Many of the previous large meanders were cut off and became dead channels. Drainage conditions were further altered by the lowering of the ground water table, although this, to a certain extent, was countered by a slight rise due to the increased rainfall of the period. Drainage changes, in turn, led to important changes in soil formation and vegetation cover. The appearance of alkali forests along the old oxbows probably dates from this period and was a direct consequence of drainage changes which followed the valley incision of the rivers.

As far as archaeological recovery is concerned, sites dating from the Atlantic period would be expected to occur above the present floodplain, on or above the Subboreal terrace. This indeed is the case in the Tiszazug region, where all the Körös and Linear Pottery sites are found above the present level of the floodplain, on the so-called old-Holocene terraces (Nagy, 1954). Some of these terraces may in fact be Pleistocene in age (Somogyi, 1967). In the Bereg-Szatmár Plain, on the other hand, valley incision does not seem to have been significant. Here the Tisza and its tributaries meander in raised beds across their alluvial cones. At Szamossályi the Linear Pottery site was found under 2 m deep alluvium (Stefánovits, 1969).

While the terrace morphology of the Danube in Hungary is relatively well understood (Pécsi, 1959), the terrace systems of many other rivers, including certain sections of the Tisza, have not been fully worked out yet. In many cases, the age of the terraces is controversial. Until the terraces are identified and more securely dated, the reconstruction of the Atlantic floodplain surface has to remain tentative, at least in some areas. It should also be remembered that from the Atlantic period onwards, man's influence has increasingly to be taken into account in explaining subsequent changes in river discharge and resultant aggradation or degradation. The downcutting in the Subboreal phase in Hungary seems however to have been the result of climatic and tectonic changes rather than the outcome of man's interference.

D. GEOLOGY AND RELIEF

Differences in rock composition and structure are important factors in determining the character of surface topography and, to some extent, the nature of the soil mantle of a region. In Hungary 8 types of metamorphic, 11 types of igneous and 60 types of sedimentary rocks have been distinguished (National Atlas of Hungary, 1967). In age the formations range from Precambrian to recent Holocene. Differences in the geological structure of the Transdanubian Central Mountains, the inselbergs (Mecsek, Villány, Bükk, Sopron and Kőszeg Mountains), the Transdanubian Hill Country and the Great and Little Plains are very marked. As a rule, areas which are topographically high are underlain by 'older' and 'harder' rocks, and those which are low by 'younger' and 'weaker' rocks.

The oldest formations in Hungary, the granite outcrops of the Velence and Mórágy Hill and the crystalline schists of the Sopron and Kőszeg Mountains, are fragments of the so-called Tisia massif, itself a remnant of the ancient European and Asian Variscan mountain system (Pécsi and Sárfalvi, 1964). The Tisia and similar, mutually isolated south European massifs played an important role in shaping the Alpine-Carpathian-Dinaric frame of the present basin.

The limestone and dolomite cover of the Hungarian Mountains was formed during the Mezozoic, when the place of the large mountain chains, the Carpathians, Alps and Dinarids, was occupied by a deep marine trough, the Tethys. The Transdanubian Mountains, having emerged in the Cretaceous, became peneplaned during the Tertiary. As they were still situated in a relatively low position, sediments denuded from the Tisia massif were periodically deposited on the surface of the Mezozoic blocks.

Hungary began to develop into the basin it is today during the Tertiary. Simultaneously with the emergence of the Carpathians, the Tisia broke up and

Map 7. Settlement Distributions and Geology

Key

HOLOCENE
- 1 Fluviatile gravel
- 2 Fluviatile sand
- 3 Wind-blown sand
- 4 Calcareous mud
- 5 Peat, muck
- 6 Meadow clay
- 7 Brown earth
- 8 Loessic silt
- 9 Alkali flats
- 10 Alluvial sand, silt, clay

PLEISTOCENE
- 11 Gravel
- 12 Fluvial sand
- 13 Wind-blown sand
- 14 Loessic sand
- 15 Sandy loess
- 16 Loess, loam
- 17 Infused loess
- 18 Clayey loess
- 19 Clay
- 20 Brown and red earths
- 21 Tufa

PLIOCENE
- 22 Basalt, basaltic tuff
- 23 Sandstone
- 24 Tufa
- 25 Clay, sand, sandstone
- 26 Sand, clay, gravel

MIOCENE
- 27 Gravel, sand, clay
- 28 Coarse textured limestone
- 29 Leithakalk
- 30 Andesite
- 31 Andesite tuff
- 32 Dacite, trachyte
- 33 Rhyolite
- 34 Rhyolite tuff
- 35 Sandy clayey marl
- 36 Conglomerate
- 37 Sand, sandstone

OLIGOCENE
- 38 Sand, sandstone, sandy clayey marl
- 39 Clay, clayey marl
- 40 Calcareous marl

EOCENE
- 41 Limestone, marl, clay
- 42 Limestone, marl, sand
- 43 Calcareous marl

CRETACEOUS
- 44 Conglomerate
- 45 Marl, limestone
- 46 Bauxite, bauxitic clay
- 47 Diabase, gabbro, wehrlite
- 48 Quartz porphyry and tuff
- 49 Trachydolerite and tuff
- 50 Marl, limestone, sandstone

JURASSIC
- 51 Siliceous marl
- 52 Limestone and marl
- 53 Mottled marl
- 54 Brachiopodal and ammonitic limestone

TRIASSIC
- 55 Sandstone, clay shale
- 56 Haupt dolomit
- 57 Upper Marl group and Dachstein kalk
- 58 Diabase, porphyrite and tuffs
- 59 Dolomite
- 60 Clay shale and sandstone
- 61 Limestone and marl
- 62 Variegated clay shale, limestone, dolomite

PERMIAN
- 63 Thick-bedded and slaty red sandstone
- 64 Grey sandstone and clay shale
- 65 Black bituminous limestone
- 66 Multicoloured slate

CARBONIFEROUS
- 67 Dark grey clay shale with limestone lenses
- 68 Dark grey clay shale and sandstone
- 69 Limestone and shale sequence
- 70 Light grey semicrystalline limestone
- 71 Granite, migmatite and contact schists

DEVONIAN
- 72 Whitish crystalline limestone and dolomite
- 73 Dark grey sericitic clay shale and sandstone
- 74 Grey-banded white crystalline limestone

EARLY PALAEOZOIC
- 75 Green schist with serpentinite
- 76 Quartz-mica schists
- 77 Calcareous mica schists

PRECAMBRIAN
- 78 Mica schist, leucophyllite, quartzite
- 79 Gneiss

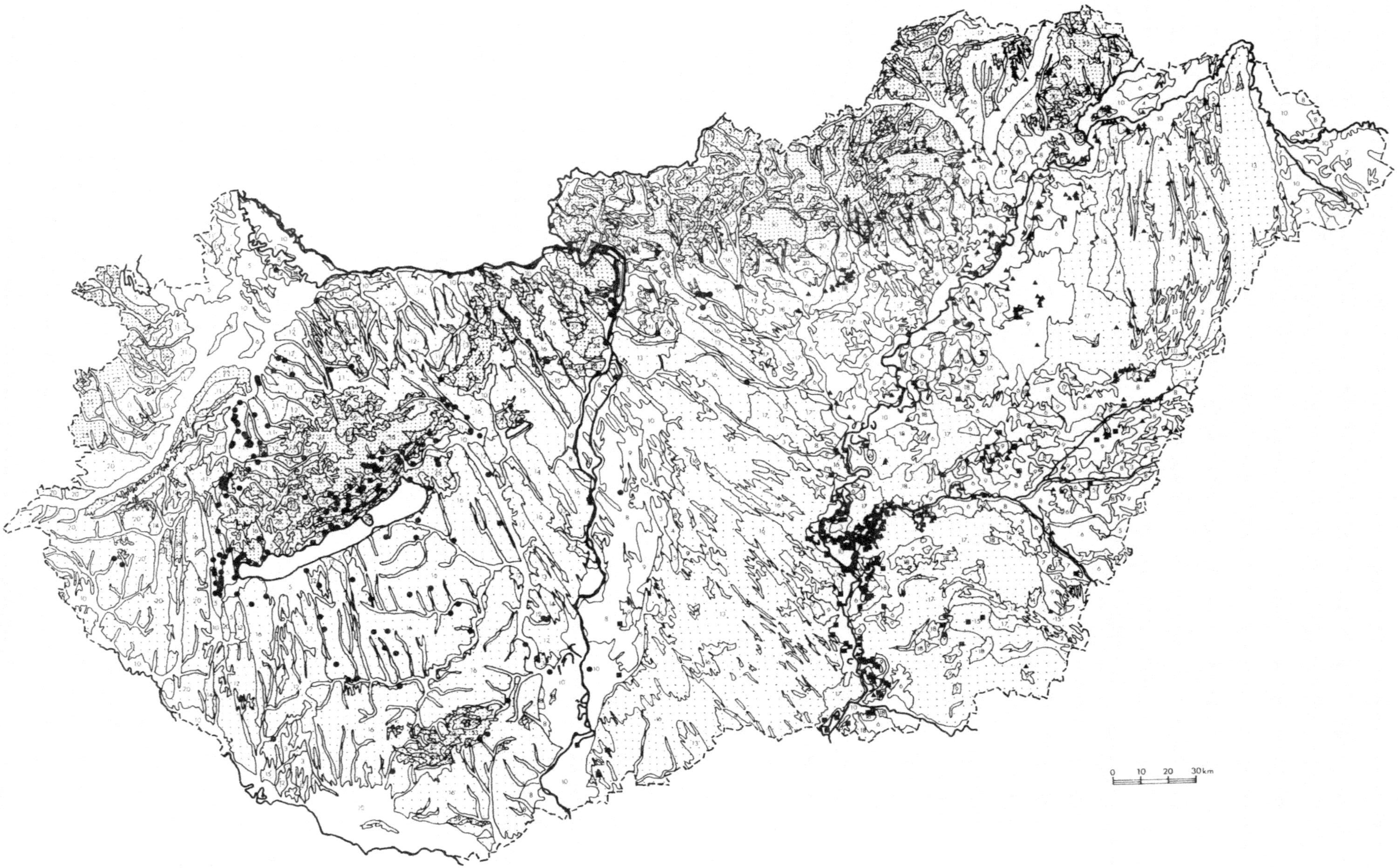
0 10 20 30km

began to subside. The resultant basin was inundated by the Miocene-Tortonian Sea. The subsidence of the Tisia was accompanied by violent volcanic activity, the result of which was the emergence of the young volcanic belt of the Börzsöny, Cserhát, Mátra and Zemplén Mountains at the edge of the basin.

In the Pliocene, thelast phase of the Tertiary, the Carpathian Basin was covered by the Pannonian Sea. Due to renewed tectonic upheavals, the Tisia subsided again and in places was covered by 2-3000 meters of thick sandy clayey Pannonian sediment. As the result of repeated crustal movements at the end of the Pliocene, the Pannonian Sea was cut off and became a fresh water lake, which was gradually silted up by fluviatile sediments. Simultaneously with the relative subsidence of the plains after the withdrawal of the Pannonian Sea, the whole of the Carpathian Basin underwent a general uplift. In the hilly country of Transdanubia and along the foothills of the Hungarian Mountains, the Pannonian sediments occur at altitudes of 2-400 meters, while in the Great Plain they are overlain by several meters of post-Pannonian deposits.

Orogenies continued in the Quaternary period. The Hungarian Mountains were again uplifted, and there was a further subsidence of the basins. The low-lying areas were filled up by several hundred meters thick fluviatile sediments. In the rising areas the rivers cut deep terraced valleys, while in the transition zone between the basins and the mountains they built vast alluvial fans. Wind action was also important in shaping the relief; during the Pleistocene extensive areas of aeolian loess and sand were formed.

The loess cover is thickest in Transdanubia, where its thickness may be as much as 30 to 40 meters. In the Great Plain, the loess consists of reworked floodplain silt, the properties of which are slightly different from those of typical loess. The loess-like sediment that covers the slopes of the Hungarian Mountains, is mainly stratified slope-loess, a product of glacial solifluction. The extensive areas of aeolian sand between the Danube and the Tisza and in the Nyirség region were also formed by wind action during the Quaternary period.

The sediments redeposited by the rivers or by the winds were mainly the products of accelerated weathering and erosion caused by the violent climatic fluctuations of the Pleistocene. A vast amount of material was redeposited at the time; about half (54%) of Hungary's present surface is covered by Pleistocene sediment.

The most recent geological formations in Hungary are the floodplain areas. These are still changing today. Pleistocene and Holocene deposits together cover more than three quarters of Hungary's surface.

Settlement Distributions and Subsurface Geology

The distribution of settlements according to subsurface geology is shown in Map 7. The percentage of sites associated with the various rocks are shown in Table 7. The results are presented in tables rather than histograms because of the large number of geological categories.

The greatest number of sites in all three cultures are found on loess or loess-like deposits. The percentage figures are: 55 in the Körös, 44 in the ALP and 42 in the TLP. The rest of the sites are on various non-loessic deposits. In the Körös culture, 40% are found on Holocene deposits, mostly alluvial in origin, and 5% on Pleistocene sand. In the ALP, 30% are on Holocene deposits, 12% on sand; 4% on volcanic tuffs; 4% on limestones and dolomites; 3% on sandstones; 1% on volcanic rocks; and less than 1% on shales. In the TLP, 12% are on Holocene alluvium, 13% on sand, 14% on Pleistocene gravel, 3% on volcanic rocks and tuffs, 8% on limestone and dolomites and 9% on sandstones.

In the Early Phase of the ALP, 41% of the sites are on loess and loess-like sediments; 31% are on Holocene alluvium; and a relatively high, 24% on sand. The remaining 3% are associated with volcanic rocks. The settlements of the Middle Phase are still relatively restricted in their distribution; 4 9% are on loess, 37% on Holocene alluvium, and 10% on sand. Of the remaining sites, 2% are on volcanic tuffs, 2% on shale and less than 1% on sandstone. In the Late Phase, the settlements are found on a much greater variety of rocks, although the highest percentage, 45, is still associated with loess. The rest of the sites show the following distribution: 27% on Holocene alluvium, 12% on sand, 6% on limestone and dolomite, 6% on volcanic tuffs, 3% on sandstone, 2% on volcanic rocks, and less than 1% on shale.

In Early TLP, 43% of the sites are on loess, 28% on alluvium, 14% on sand, and 14% on sandstone. In the Middle Phase, 42% are on loess, 8% on alluvium, 16% on Pleistocene gravel, 11% on sand, 10% on sandstone, 9% on limestone and dolomite, 4% on volcanic tuffs. In the Late Phase,38% are on loess, 19 on alluvium, 8% on gravel, 16% on sand, 5% on sandstone, 11% on limestone and dolomite, and 3% on volcanic tuffs.

The late groups of the ALP show some variation in their distributions. In the Esztár and Szilmeg groups the highest percentage of sites are on Holocene alluvium. The second highest percentage in the Esztár group is associated with sand. In all the other groups, the highest number of sites is found on loess. With the exception of the Szakálhát group, all the other groups have settlements on volcanic tuffs. Otherwise, the distributions of the Esztár, Szilmeg and Szakálhát groups are relatively limited. In contrast, the settlements of Late AVK, Tiszadob and Bükk are found on a much greater variety of rocks. Especially noticeable is the relatively high percentage of sites on limestone in Late AVK, Tiszadob and Bükk; and on rhyolite tuff in Tiszadob, Bükk and Szilmeg groups. The increased interest of the Late Phase sites in sandy regions has already been noted.

Chi-square tests suggest that the probability of chance distribution in either culture is less than 0.05%. Percentage point difference between the expected and observed distributions are shown in Table 7. All the values are relatively high: they range from 41.24 to 67.92. The lowest values are again found in the ALP, where chronological and regional differences are fairly strong.

The importance of the loess areas in the spread of the Linear Pottery culture in Central Europe is well known and has been often emphasised by archaeologists (Childe, 1929; Clark, 1952; Butzer, 1971). One attraction for settlement was the natural vegetation cover of the loess. In the Atlantic period loess areas carried forest-steppe in the drier and mixed oak forest associations in the wetter areas (Firbas, 1949-52). Both the forest-steppe and the mixed oak forest provided plentiful fodder for domesticated animals.

Table 7 Percentage point difference between expected and observed distributions of settlements in relation to subsurface geology

Geology	Kőrős culture			ALP		ALP	
	Expected %	Observed %	Ppd	Observed %	Ppd	Observed %	Ppd
1	0.16						
2	1.01						
3	2.17			1.04		1.42	
4	0.68						
5	1.49	0.77					
6	3.50			1.04			
7	0.90						
8	3.80	9.92	-6.12	9.09	-5.29		
9	4.73	9.92	-5.19	8.31	-3.58		
10	12.36	19.08	-6.72	11.17		10.85	
11	2.47					13.68	-11.21
12	4.18					8.96	- 4.78
13	11.95	5.34		10.65		2.36	
14	3.37	0.77		0.26		1.42	
15	2.01	2.29	-0.28	3.64	-1.63	2.36	-0.35
16	15.49	0.77		4.93		37.26	-21.77
17	9.07	41.98	-32.91	28.83	-19.76		
18	1.52	9.16	-7.64	5.19	-3.67		
20	4.05			2.60		0.47	
22	0.24					0.47	-0.23
23	0.16						
24	0.08					0.47	-0.39
25	3.34			0.52		7.08	-3.74
26	0.16						
27	0.63			0.26		0.47	
28	0.44						
29	0.41		0.26	0.26			
30	1.06			0.78			
31	0.79					0.47	
33	0.24			0.52	-0.28		
34	0.41			4.41	-4.		
35	0.52						
36	0.22					0.94	-0.72
37	0.92			2.08		-1.16	
38	1.77						
39	0.27					0.47	-0.20
40	0.19						
41	0.16			0.26	-0.10		
43	0.11						
44	0.03						
45	0.03						
47	0.03						

(contd.)

Table 7 (contd.)

Geology	Kõrös culture Expected %	Observed %	Ppd	ALP Observed %	Ppd	TLP Observed %	Ppd
47	0.03						
49	0.08						
50	0.11						
51	0.03						
53	0.05						
54	0.05						
55	0.03						
56	0.95					1.89	-0.94
57	0.35					3.77	-3.42
58	0.11					2.36	-2.25
59	0.27			0.26			
60	0.14			0.26	-0.12		
61	0.30			3.12	-2.82		
62	0.16					1.42	-1.26
63	0.05					0.47	-0.42
66	0.03						
67	0.03			0.52	-0.49		
73	0.05						
75	0.03						
77	0.03						
79	0.03						
Total Ppd			58.86		42.90		51.68

Ppd = Percentage point difference

Geology	ALP Early Observed %	Ppd	Middle Observed %	Ppd	Late Observed %	Ppd
3	6.90	-4.73	0.79		0.40	
6			0.79		1.19	
8	13.79	-9.99	11.11	-7.31	7.93	-4.13
9	6.90	-2.17	12.70	-7.97	6.75	-2.02
10	10.34		11.91		10.71	
13	17.24	-5.29	8.73		11.11	
14	3.45	-0.08				
15			5.56	-3.55	2.38	-0.37
16			3.17		6.35	
17	31.03	-21.96	29.37	-20.30	28.97	-19.90
18			8.73	-7.21	4.76	-3.24
20	6.90		2.38		2.78	
25					0.79	

(contd.)

Table 7 (contd.)

ALP

	Early		Middle		Late	
	Observed %	Ppd	Observed %	Ppd	Observed %	Ppd
27					0.40	
29					0.40	
30					1.19	-0.13
33	3.45	-3.21			0.40	-0.16
34			2.38	-1.97	5.55	-5.14
37			0.79		1.98	-1.06
41					0.40	-0.24
59					0.40	-0.13
60					0.40	-0.26
61					4.76	-4.46
67				1.59	-1.56	
Total Ppd		47.43		49.87		41.24

TLP

	Early		Middle		Late	
3					3.17	-1.
5			1.26			
10	28.58	-16.22	6.92		19.05	-6.69
111			15.72	-13.25	7.93	-5.46
12	14.28	-10.10	10.06	-5.88	6.35	-2.17
13			0.63		6.35	
14			3.14		1.59	
15			1.26		4.76	-2.75
16	42.86	-27.37	37.11	-21.62	31.74	-16.25
20			0.63			
22			0.63	-0.39		
24			0.63	-0.55		
25			8.18	-4.84	3.17	
27			0.63		1.59	-0.96
31					1.59	-0.80
36			1.26	-1.04		
39					1.59	-1.32
56			2.51	-1.56	1.59	-0.64
57			4.40	-4.05	6.35	-6.
58			3.14	-3.03	1.59	-1.48
62			1.89	-1.73	1.59	-1.43
63	14.28	-14.23				
Total Ppd		67.92		57.94		46.95

(contd.)

Table 7 (contd.)

Geology	Late ALP					
	Late AVK		Tiszadob		Esztár	
	Observed %	Ppd	Observed %	Ppd	Observed %	Ppd
3					2.78	-0.61
8	6.18	-2.38	2.49		30.56	-26.76
9	7.41	-2.68	7.46	-2.73	8.33	-3.60
10	9.88		2.99		8.33	
13	4.94		16.42	-4.47	25.	-13.05
15	2.47	-0.46	1.49		8.33	-6.32
16	7.41		16.42	-0.93		
17	25.93	-16.85	17.91	-8.84	13.89	-4.82
18	7.41	-5.89				
20	3.70		5.97	-1.92		
25	1.23		1.49			
29	1.23	-0.82				
30	2.47	-1.41	2.99	-1.93		
33	1.23	-0.99	1.49	-1.25		
34	4.94	-4.53	8.95	-8.54	2.78	-2.37
37	3.70	-2.78	4.48	-3.56		
59	1.23	-0.96				
61	8.64	-8.34	8.95	-8.65		
Total Ppd		48.09		42.82		57.53

	Late ALP					
	Szilmeg		Bükk		Szakálhát	
6	3.57	-0.07			3.66	-0.16
8	14.28	-10.48	1.28		8.54	-4.74
9	17.86	-13.13	8.97	-4.24	7.32	-2.59
10	3.57		6.41		14.63	-2.27
13	17.86	-5.91	15.38	-3.43		
15	3.57	-1.56				
16			12.82		2.44	
17	21.44	-12.37	16.67	-7.60	50.	-40.93
18	3.57	-2.05			10.97	-9.45
20	3.57		2.57		2.44	
27			1.28	-0.65		
29			1.28	-0.87		
30			2.57	-1.51		
34	10.71	-10.30	10.26	-9.85		
37			3.85	-2.93		
41			1.28	-1.12		
59			1.28	-1.01		
60			1.28	-1.14		
61			12.82	-12.52		
Total Ppd		55.87		46.87		60.14

The lightly stocked oak forest, even before the intervention of man, admitted sufficient light to allow the growth of a rich underlayer and thus provided adequate feed for cattle and swine (Clark, 1952).

The main reasons for the occupation of the loess areas, however, were high soil fertility, good drainage and ease of working. 'Loessical parent materials contain large amounts of silt, about 10 to 20% clay, have a high content of weatherable minerals, show high base saturation, or are calcareous. As these parent materials are on relatively young landscapes in a mild climate, they have been only slightly modified by pedochemical weathering. Consequently, soils formed from these initial materials are silty, high in nutrient reserve and generally have excellent physical properties' (Buol, Hole and McCracken, 1973, 110). Not all loess areas are equally fertile, however, or offer the same advantages for settlement. In the northern Rhineland, for example, the Bandkeramik sites are located in the southern part of the loess region, where the soils have a higher degree of basal saturation, while the northern loess are devoid of settlement (Narr, 1956).

In Hungary, only the TLP sites are found on typical loess. Most of the Körös and ALP sites are found on a different loess type, the so-called 'infusion' loess. Infusion loess is really a transitional type, which combines the features of real loess and river alluvium. It was formed on the old floodplains of the Alföld from loessic silt redeposited by river action. If examined in section, only its uppermost 1.5-2 meter part exhibits the characteristic features of true loess (Pécsi, 1967). Below that it consists of a succession of alluvial sand, silt and clay layers. The buried soil horizons are usually of alluvial or meadow soils. The silt content of the parent material is usually high and drainage conditions are greatly influenced by microtopography or the position of an impermeable layer. In contrast with Transdanubia, loessic parent material in the Alföld is not necessarily a guarantee of good drainage or a salt-free horizon. When the water economy is favourable, on the other hand, soils developed on infusion loess can be very fertile. The fertility is enhanced by the fact that the ground water table is located at a depth within reach of agricultural plants (Somogyi, 1971). Soil conditions can change rapidly, however, within a short distance.

The loess cover of Transdanubia is composed of typical or ordinary loess, which is the most widespread loess type in western Europe. The most fertile soils are found on the southern slopes or foothills of the hills and mountains. The loess layer in this position usually contains several old buried soil horizons, the nutrient content of which, especially when mixed with volcanic weathering products,is very high (Pécsi, 1967). Erosion due to the greater cohesion of the old surfaces is usually less of a hazard than in other loess types. Today, the best vine and fruit orchards are found on this parent material. Many of the Linear Pottery sites are also located near or on such formations.

The 40-50% figures for the occupation of the loess areas in Hungary are relatively low when compared with Germany, where on the average 80% of the Bandkeramik sites are found on loess (Sielmann, 1972). The low values are, however, somewhat deceptive. Holocene terraces, levees and low-lying basins contain an appreciable amount of loess, even though these surfaces are not counted as loess areas (Balogh, et al., 1958). It has also to be taken into account that some of the smaller loess areas had to be omitted from the map because of scale (Marosi, 1954). In spite of these qualifications and the undoubted preference of the early agriculturalists for the loess areas in mid-latitude Europe, the association with loess is not exclusive. Settlements also occur with some frequency on non-loessic deposits.

A number of ALP and TLP sites are located on volcanic rocks and tuffs, limestones, dolomites and sandstones. As yet no Körös site has been found on other than recent Quaternary fluviatile or aeolian deposits. The importance of the older rocks was partly to provide raw materials for the manufacturing of tools.

One of the most sought after raw materials of the pre-metal age was obsidian. The largest obsidian deposits in southeast and central Europe are in the Tokaj and Hegyalja Mountains of Hungary. The obsidian here is usually found in association with rhyolite, mostly with rhyolite tuff (Varga, pers. comm.). Small quantities of obsidian may also have been present in other rhyolite tuff outcrops along the foothills of the Bükk and Matra Mountains, even if today there is hardly any obsidian there. It is probably no coincidence that obsidian tools constitute a relatively small part of the tool assemblage of the Körös culture, but are very numerous in the ALP (Csalog, 1955). According to Kalicz (pers. comm.),80% of the tool kit of Méhtelek is made of obsidian and at Tiszacsege, an Early ALP site, only obsidian tools have been found. The number of sites on rhyolite tuff increases considerably in the Late Phase. This is also the period when there is some evidence for the existence of settlements specialised in the manufacture of obsidian and flint tools (Kalicz and Makkay, 1977).

Although nearness to a major obsidian source was undoubtedly an important factor in the location of certain ALP settlements, it should not be forgotten that soils formed on volcanic tuffs can be very fertile. Volcanic soils, even without obsidian,must have been attractive to the early agriculturalists. In the Middle Rhine region of Germany, the largest number of Bandkeramik settlements after loess are located on volcanic tuffs (Sielmann, 1972).

Flint and chert frequently used for tools are found mostly in Mezozoic limestones in Hungary (Varga, pers. comm.). Beside being sources of raw materials, limestones are also preferred by some domestic animals; sheep and goats, if allowed to graze freely, move to limestone soils and vegetation in preference to all other soils (Hughes, 1958). In the Bükk Mountains there is a number of limestone caves which were occupied through the duration of the ALP. It is not yet known whether the caves were occupied all the year round or used only on a seasonal basis. In the cold winter mountain region, where good agricultural land is relatively scarce, it is not unlikely that transhumance was an established practice. Conclusive evidence, however, is still missing.

The main purpose of the TLP settlements established on the dolomite blocks of Transdanubia may also have been the pasturing of animals and the procurement of raw materials, although the climate here is more favourable to all the year occupation. There are also small pockets of good if limited agricultural land in the vicinity of the sites. In both areas further study is necessary to determine the most likely function of the settlements.

Sandstones seem to have been more frequently occupied by TLP than ALP settlements. The Upper Pannonian sandstone on which the settlements are most frequently found does not occur in pure deposits; it is usually found intercalcated between layers of clay, sand, gravel and marl (Balogh et al., 1958). The sandstone itself is a useful material for the production of neolithic axes, while the soils developed on the mixed clay, sand and sandstone deposits are generally well drained and relatively rich in nutrients.

While relatively few Körös sites are located on sand, over 10% of the settlements in the ALP and TLP cultures are found in sandy areas. Sandy soils are usually well drained but often low in fertility. As will be shown in the next section, areas of very low fertility were avoided by the Linear Pottery settlements, and only sandy soils of relatively high fertility were occupied.

Over 13% of the TLP sites are located on Pleistocene gravel. Most of the settlements are in the Marcal region on terraces cut into previous alluvial fans. The gravel here is usually mixed with sand and loess and provides very good drainage. It is of interest that the settlements are restricted to those areas which are underlain by calcareous gravels (Goczán, 1971). The soil cover of the Marcal region will be discussed later in greater detail.

The geological structure of Hungary has changed relatively little since the Atlantic period. The greatest changes undoubtedly occurred in the alluvial areas of the rivers. The development of the floodplain regions since the Atlantic has already been described. The other changes that occurred were mostly brought about by man. Man-induced erosion, for example, has been especially strong in Transdanubia. The loess areas which have attracted agriculturalists from an early time onwards, are very vulnerable to erosion. It is unlikely, however, that erosion would have been too advanced at the time of the Linear Pottery occupation.

Settlement Distributions and Relief

Hungary's topography is characterised by the predominance of low-lying plains, gently rolling hills and by a total lack of high mountains. Only 2% of the total area of the country rises 400 metres above sea level; 14% is hilly country between 200 to 400 metres in elevation; and 84% is low-lying level and hilly terrain (National Atlas of Hungary, 1967).

The distribution of settlements in relation to relief is shown in Map 1. The settlements of the Körös culture are all located in the Great Plain. Almost all the sites are in areas which are below 100 m s.l. Most of the ALP and TLP sites are also located in areas of low elevation, in level plains or low foothills.

The sites of Mencshely-Murvagödrök, Vöröstó-Ragonya and Hidegkút in the TLP are exceptional in being at 300 m above sea level. The reason for the location of the sites at Mencshely and Vöröstó is probably the nearby presence of easily accessible flint and quarzite deposits. Surface material found at the sites suggests that mining and manufacturing were carried out by the settlers.

In Late ALP, especially in the later Bükk phases, there is a tendency to locate settlements in new,previously avoided areas. The occupation of cave sites increases and settlements appear on high, isolated hilltops. The site of Boldogkőváralja-Leányvár is at 367 m a.s.l. and Tolcsva-Várhegy at 485 m. Both sites could easily be defended. The reason for the location of the new sites are not yet clearly understood. It is possible that intersite economic specialisation was one reason, and a growing population another; other reasons can also be postulated. Future research will probably provide the answer.

E. SOILS

Soils are, next to water, man's most important natural resource. As the medium from which plants grow, soils provide food for man and animals. The influence of soil variation on the subsistence base of non-agricultural societies is mainly indirect, while in agricultural economies soil fertility and productivity are naturally of primary importance.

Soil formation is generally viewed as dependent on five major factors. According to Jenny's (1941) fundamental equation,

$$\underline{S} \text{ or } \underline{s} = f\,(\underline{cl},\ \underline{o},\ \underline{r},\ \underline{p},\ \underline{t})$$

where $\underline{S}$ denotes the soil, $\underline{s}$ any soil property, $\underline{cl}$ climate, $\underline{o}$ biota or organisms, $\underline{r}$ topography or relief (including ground water), $\underline{p}$ parent material, and $\underline{t}$ time or duration of soil formation. In the equation, the five soil-forming factors are regarded as independent variables, with soil or soil properties the dependent variable. Although the independence of the five factors is more theoretical than actual, the equation does emphasise that the relationship between soil and the soil-forming factors is a functional one: a change in any of the five variables will bring a corresponding change in the dependent variable. Because of the functional relationship between soil and the soil forming factors, the analysis of settlement distribution according to soil type is also, to some extent, a synthesis of the climatic, biotic, hydrographic, and geological conditions at the sites.

The basis for the following analysis of settlement distributions is the soil map of Hungary (National Atlas, 1967). Hungarian soil classification, like the Russian, has a strong genetic emphasis; soil properties and pedogenic processes are evaluated in relation to soil-forming factors. The major unit of classification is the soil type, which is defined as having 'unity of origin, substance transformation, migration and accumulation; that is, they are generalisations of many actual soils which have a common origin and similar pedogenic processes' (Buol, Hole and McCracken, 1973). Because of the intermediate character of Hungary's climate and the mosaic-like nature of the hydrographic, geological and relief features, the number of soil types in spite of the relatively small size of the country is quite large (Pécsi and Góczán, 1971). On the most recent soil map 29 soil types and subtypes have been distinguished. The 29 types and subtypes form 7 major soil groups: 1) skeletal soils; 2) lithomorphous soils; 3) brown forest soils; 4) chernozem soils; 5) alkali soils; 6) meadow and floodplain forest soils; and 7) bog soils.

The forest and chernozem soils in Hungary are zonal; their characteristics are mainly determined by climate. Azonal are the skeletal soils, which have no horizon differentiation, while the rest of the soil groups are termed as intrazonal. Intrazonal soils, in spite of climate and vegetation, reflect the influence of some local condition such as poor drainage or alkalinity (Pécsi and Góczán, 1971; Buckman and Brady, 1969). Before attempting to evaluate the effect of soil variation on the distribution of the Körös and Linear Pottery settlements, the main soil types of Hungary and their potential for successful crop production are described briefly.

The Soil Groups of Hungary

1. Skeletal soils: Blown sands and alluvial soils are the two main soil types in this group.

Blown sands occupy large areas in the Great Plain. They contain little humus and have poorly developed structure. The sands between the Danube and the Tisza are calcareous, while those in the Nyirség are slightly acid. Unstabilised sands are of little agricultural value.

Alluvial soils which are the result of deposition by rivers are also very extensive in Hungary, especially in the Little and Great Plain. These soils contain very little humus as soil formation is retarded by the often recurring floods. On the higher floodplains which are reached only by occasional floods, humic alluvial soils have developed. These in the course of time are transformed into alluvial chernozems or alluvial meadow soils. Alluvial soils can be very fertile, but the threat of repeated flooding is sometimes a strong deterrent to regular agricultural utilisation.

2. Lithomorphous soils: The characteristics of lithomorphous soils are to a large extent determined by the nature of the parent material. In Hungary the two main soil types that belong to this group are the rendzinas and the so-called 'nyirok' or erubase soils.

Rendzinas are rather thin grassland soils developed from soft limestone rocks. In Hungary they are found on the limestone and dolomite plateaus and slopes of the Hungarian Mountains under rock grasses and karst shrub forests or occasionally under hairy oak forests. As they dry out quickly, their agricultural productivity is low.

Erubase soils are formed from lavas and volcanic tuffs. They usually occur under extreme soil-climate conditions, on steep ridges and dry southern slopes. These soils have a high clay content and are usually very high in humus. Their reaction is slightly basic or neutral. In spite of the high humus content, the productivity of erubase soils is low, due to the extreme fluctuation of the microclimate.

3. Brown forest soils: About 36% of Hungary's surface is covered by brown forest soils (Stefánovits, 1971). Forest soils, as the name indicates, are formed in areas which have adequate precipitation for forest growth. The major processes that operate in the formation of forest soils are leaching, podzolisation and lessivage. Leaching is the washing out of the soluble materials of the solum. In some forest soils, leaching is only sufficient to remove the calcium carbonate and does not seriously impair the natural fertility of the soil. In others, leaching may be so severe and consequently the upper layers so acidic that agricultural production becomes marginal (Buckman and Brady, 1969).

Lessivage is the mechanical migration of clay particles from the A to the B horizons. Podzolisation is the chemical migration of aluminium and iron and/or organic matter resulting in the concentration of silica in the eluvial horizon (Buol, Hole and McCracken, 1973). Podzolisation and lessivage may or may not operate simultaneously with the leaching process.

In Hungary, the main soil types of the brown forest group are: strongly acid non-podzolic brown forest soil; podzolic brown forest soil, brown forest soil with clay illuvium (lessivated br. forest soil), pseudogely type brown forest soil; Ramann's brown earth, 'kovárvány' brown forest soil;and chernozem-brown forest soil.

Acid non-podzolic brown forest soils: These soils are formed on argillites, phyllites, porphyrites and hydroandezites. The leached horizon is rich in organic matter, mainly because the ground vegetation consists of heather and cranberry; the humus type is acid mull. The pH values of this horizon are 3.5-4.5. In the illuvial horizon, acidity is also significant; pH is around 5.

The moisture regime is unfavourable. As the soils are mostly shallow and have low water holding capacity, they dry out quickly in summer. The nutrient status is low; they contain little nitrogen, phosphorus and potassium. In western Europe these soils are referred to as 'sols bruns acids'.

Podzolic brown forest soils (podzolic lessivated br. f. soils): These soils are found under closed forests in areas of high precipitation. They are especially characteristic of the higher regions of the Northern Mountains, where they mostly occur under humid oak and beech associations. They are also found under pine forests in the foothills of the Alps.

The reaction of the whole soil profile is acidic. The humus layer is very thin. Nutrients are in short supply; nitrogen deficiency is pronounced. Plants sometimes suffer from drought in summer, as the moisture regime of the A horizon fluctuates greatly in spring and summer.

Brown forest soils with clay illuvium (Lessivated br. f. soils): Of the forest soils these are the most extensive in Hungary. They are found in mountainous and hilly regions, where the annual precipitation is over 650 mm. They usually occur under acidiphilous oak or hornbeam-oak or occasionally under beech forests.

Their reaction is moderately acid. The moisture regime is good. The soils have good water holding capacity and the available water is satisfactory for crop cultivation. The nutrient content is generally moderate. The supply of nitrogen is largely dependent on humus content and consequently is generally poor. Phosphate is in moderate supply, and potassium is adequate. On the whole these soils are of moderate fertility.

Pseudogley brown forest soils: In Hungary pseudogley soils are closely related to brown forest soils, although this is not necessarily the case in other regions outside Hungary. These soils occur under beech or hornbeam-oak forests or grassy meadows with plants of high moisture requirement.

The soils are strongly acidic, with pH values seldom exceeding 6.0. The physical characteristics, especially the water regime, are very unfavourable to crop production. The illuvial horizon and the other subsurface layers have a tendency to swell and contract; for most of the year the horizon is in an expanded state and this makes it compact and low in pore space. The nutrient status is also unfavourable; nitrogen and phosphates are in short supply, and potassium is often leached out.

Ramann's brown earth: Besides the chernozem-brown forest soils these are the most fertile forest soils in Hungary. Braunerden occur only in the drier parts of the country, in particular within basins and on slopes in the northern part of the brown forest soil region and in areas adjacent to the chernozem region. 'In general, where the topography is varied, lessivated brown forest soils occur in the plains and braunerden on the slopes' (Stefánovits, 1971). In some soil classifications, braunerden are termed as intrazonal soils because of the high lime content of the parent material (Buckman and Brady, 1969). In Hungary the most common parent material is loess. When not under cultivation, the vegetation cover consists of calciphile oak or beech forests.

The A horizon of braunerden has a good, crumbly or granular, structure. The moisture regime is good. The soil is moderately permeable and has a good water holding capacity, the whole profile providing water for plant growth. The reaction of the soil is slightly acid or neutral. The humus content under forest can be as high as 8%, but the same soil under arable cultivation has only 2%. Nutrients are in good supply in non-eroded profiles. Braunerden are easy to cultivate and yield rich crops, especially when fertilised occasionally.

Kovárvány brown forest soils: When brown forest soils develop on sand, the type generally formed in Hungary is 'kovárvány' brown forest soil. Kovárvány are thin clay lamellae in the accumulation horizon.

The water regime of these soils is much more favourable than is usually the case with sandy soils. The infiltration rate of the water is slowed down by the clay lamellae and plants have a longer time to absorb the moisture. Also, there is usually more water stored between the clay bands than in other sandy soils. The nutrient status is quite high, partly because of the presence of the clay lamellae, which can bind larger quantities of plant nutrients and partly because there is more time for the plants to absorb the nutrients. The fertility of kovárvány brown forest soils is moderately high.

Chernozem-brown forest soils: These soils are predominantly found in the zone of transition between plains and hilly or mountainous areas. They have certain characteristics which are typical of brown forest and others typical of chernozem soils. 'The soil profile has a rich deeply-developed humus horizon which very often reaches well into the illuvial horizon of the brown forest soil, masking its original colour and properties. In the lower part of the humus horizon, however, the structure is typical of the brown forest soils, blockly or even prismatic' (Stefánovits, 1971, 99). Besides leaching, intensive humification, a characteristic process in chernozems, has also been influential in the formation of chernozem-brown forest soils.

The moisture regime is good, as moderate permeability is accompanied by good water holding capacity. Nutrients are in good supply, nitrogen, phosphorus and potassium being adequate for most crops. Acidity is slight; the pH is rarely less than 6.5.

The genesis of these soils is somewhat controversial but it is generally thought that anthropogenic influence was probably responsible for the development of chernozem characteristics.

4. <u>Chernozem soils</u>: Chernozem soils occur in areas of lower rainfall usually under grassy vegetation. The rainfall is insufficient to leach out much of the calcium and other divalent cations. Consequently, these soils have high base saturation and are mildly acidic to basic in reaction. The soluble constituents leached from the surface horizons are not removed from the profile; they are only moved to the lower horizons. The zone of calcium carbonate accumulation occurs at the depth of the average penetration of the rain waters.

The vegetation has also contributed to the soil-forming process. Grasses are very effective in returning bases to the surface of the soil. The extensive root systems of these plants also supply large amounts of organic matter. The humus that develops stabilises the soil colloids when adequate calcium is present. This prevents the downward movement of the finer minerals, a process characteristic of forest soils (Buckman and Brady, 1969).

In Hungary chernozem soils occur in the Alföld and the Mezöföld. They are always found on parent materials which contain calcium carbonate. Most of the typical chernozems are found on loess. The depth of the ground water table is always below 5 m.

Chernozems with carbonate coating have good crumbly structure,and the humus layer can reach 150 cm in thickness. These soils are well supplied with mineral nutrients as well as nitrogen and sulfur. Typical chernozems are one of the most fertile soil types in Hungary. Because of the low and uncertain rainfall, however, their productivity is not consistently as high as that of the meadow chernozems.

Meadow chernozems are formed within the chernozem area where the capillary zone of the ground water table is in contact with the soil profile. Meadow chernozems are the most fertile soils in Hungary.

Chernozems with alkali subsoils are fairly frequent in Hungary. The surface horizons are similar to those of typical chernozems, but the alkali subsoil is more compact and less permeable. In dry years, plants will tend to suffer from drought. The fertility of the soil is largely determined by the depth of the salt free upper horizons (Szabolcs, 1966).

5. <u>Alkali soils</u>: 'Soils developed under imperfect drainage conditions in arid regions are often characterised by abnormal salt concentrations in the upper horizons. Because of the effects of salts on their properties, these soils are termed halomorphic or alkali soils' (Buckman and Brady, 1969).

Although Hungary is situated in the temperate zone and its precipitation is characterised by a summer half-year maximum, summers, especially in the southeastern part of the country, are arid. It has been observed that alkali soils in Hungary are strictly confined to areas where the rainfall in the arid half-year is less than 350–400 mm (Somogyi, 1964). Alkali soils are most widespread in the Great Plain, where they are found in association with chernozem and meadow soils. Topographically, they are usually situated below the chernozems and above the meadow soils. On the basis of differences in structure and salt content 3 main types of alkali soils have been distinguished in Hungary.

Solonchak soils occur where the salty ground water table lies at least intermittently higher than 0.5 m. The accumulation of soluble salts of sodium, calcium, magnesium and potassium occurs in the uppermost level; the salt content is usually between 0.3-0.5%. The soil profile is usually unstratified, and the horizons can be distinguished only with difficulty. Solonchak soils in Hungary are not very widespread; they are mostly found between the Danube and Tisza. Because of impeded drainage and high salt content, solonchak soils are of poor quality as croplands. Today, they are mostly utilised as meadows (Pécsi and Góczán, 1971).

Solonchak-solonetz soils usually occur where the ground water table is at a depth of 1 m. The soil profile is similar to that of the solonchaks, except there is a more developed B horizon. The reaction of the upper horizon is slightly basic due to the high concentration of alkali salts in the B2 horizon. The permeability of the solonchak-solonetz soils is very low. The usefulness of these soils for agricultural production is limited; rice is the only crop that can be grown on them with any success.

Solonetz soils are the most widespread alkali soils in Hungary. They are found where the salty ground water table is below 1 m but above 2 metres in depth. Solonetz soils are characterised by a high concentration of sodium in association with the soil colloids. The soil profile has a typical columnar structure. Salt concentration is highest in the B1 horizon. As the A horizon contains fewer water soluble salts and also has a better water holding capacity, the fertility of the solonetz soils is usually determined by the depth of the A horizon. In the best, so-called deep solonetzes, the depth of the A horizon is over 15 cm (Szabolcs, 1961).

Deep or medium solonetzes, also referred to as steppe solonetzes constitute one subtype of solonetz soils. The other subtype is the meadow solonetz. Topographically, meadow solonetzes are usually found in the lowest areas of the terrain, while steppe solonetzes usually occur higher up, in an intermediate position between meadow soils and meadow chernozems.

Soils studies of alkali soils suggest that the development of solonetz soils has been a very complex process. The effect of the river regulations on the formation of the solonetz soils is still not fully understood. Generally, solonetz soils occur in areas that had been seasonally inundated before the regulation. In the opinion of the pedologists, most of these soils had a solonetz profile even before the regulation (Máté, 1955; Szabolcs, 1961). Some of the solonetz soils, however, may have developed from solonchaks through leaching due to the general lowering of the ground water table (Szabolcs, 1961). In addition, some secondary salinisation did occur after the regulations. The question of soil changes, or at least certain aspects of it that are directly relevant to settlement distribution, will be discussed later in this chapter.

6. Meadow soils: Meadow soils are generally found in areas of excessive moisture and restricted drainage. The poor drainage is usually due to location rather than clay pan development (Buckman and Brady, 1969).

In Hungary, meadow soils are extensive on the former higher flood-plains of the rivers. They also frequently occur in silted up river beds, in larger depressions between the Danube and Tisza, and among the sand dunes of the Nyirség; that is, wherever the ground water table is near the surface. The native vegetation of meadow soils consists mostly of grasses and sedges,but they also occur under floodplain forests. Meadow soils are rich in organic matter and have thick humus layers. The calcium carbonate is usually leached out of the upper horizon, and the reaction of the soil is slightly acidic. Although they may occur on sand or loam, most meadow soils are found on heavy clays. Because of the small amount of available water in the clay, plants often suffer from drought in drier years (Nagy, 1956). The utilisation of the heavy meadow soils for agriculture usually requires the use of heavy machinery. The most frequent subtypes of meadow soils in Hungary are the meadow soils with alkali subsoil and meadow soil of steppe character.

7. Bog soils: Similarly to the meadow soils, bog soils are also characterised by an excess of soil moisture and poor drainage. In bog soils,the ground water table is at or very near the surface during most of the year.

In Hungary, bog soils are most frequently found along lake shores and in cut-off oxbows in the floodplains of the largest rivers. The high organic matter content of these soils comes from the decomposition of sedges and other swampy plant associations. According to the degree of humification of the organic matter, two types of bog soils can be distinguished: mucks and peats.

In mucks, the organic constituents are completely humified. The soil structure is very loose, light and friable. The largest muck soil areas are located east of the Tisza, between the Berettyó and Körös rivers, and in the Rétköz. Muck soils are also found in the western part of the country, in the Hanság, Little Balaton, and Sárrét.

Peaty soils are found in the same regions as the muck soils. In peats, only the upper layer is humified; the lower layers are composed of still recognisable plant remains. Peats are mostly utilised as fuel or as fertilisers.

Settlement Distributions and Soil Types

The distribution of settlements according to soil type is shown on Map 8. The percentages of Körös, ALP and TLP sites associated with the various soil types are shown in Histograms a, b, and c (Fig. 11). Histograms d, e, f, h, i and j (Fig. 11) show the percentage relationships in the Early, Middle and Late phases of the ALP and TLP. Histograms k, l, m, n, o and p (Fig. 11) give the percentage figures for the regional groups of the Late ALP.

The majority of Körös sites are found on soil types where water has been an important soil forming factor. The highest percentage (44) of settlements are on alluvial meadow soils; 13% are on alkali soils and 12% on hydromorphic soils. Of the rest, a relatively high percentage, 27, are on meadow chernozems. Meadow chernozems are, as we have seen, transitional between

MAP 8. Settlement Distributions and Soil Types

Key

1 Wind-blown sand
2 Rendzina
3 Erubase soil, black 'nyirok'
4 Strongly acid non-podzolic brown forest soil
5 Podzolic brown forest soil
6 Brown forest soil with clay illuvium
7 Pseudogley-type brown forest soil
8 Brown earth, Ramann's brown forest soil
9 Brown forest soil alternating with thin layers of clay (kovárvány)
10 Chernozem-brown forest soil
11 Chernozem-type sand
12 Typical chernozem with carbonate films (pseudomycelium)
13 Lowland chernozem with carbonate films (pseudomycelium)
14 Lowland chernozem salty in its deeper horizons, with carbonate films (pseudomycelium)
15 Meadow chernozem
16 Meadow chernozem salty in its deeper horizons
17 Alluvial chernozem
18 Solonchak
19 Solonchak-solonetz
20 Meadow solonetz
21 Meadow solonetz in transition to steppe soil
22 Solonetz-like meadow soil
23 Meadow soil
24 Alluvial meadow soil
25 Marshy meadow soil
26 Lowmoor, fensoil
27 Drained and cultivated lowmoor (fen) soil
28 Soils of swamp forests
29 Recent raw alluvial soil

0 10 20 30km

Fig. 11 Histograms showing the distribution of settlements in relation to soil types

1, wind-blown sand
2, rendzina
3, erubase soil, black 'nyirok'
4, strongly acid non-podzolic brown forest soil
5, podzolic brown forest soil
6, brown forest soil with clay illuvium
7, pseudogley-type brown forest soil
8, brown earth, Ramann's brown forest soil
9, kovárvány brown forest soil
10, chernozem-brown forest soil
11, chernozem-type sand
12, typical chernozem with carbonate films
13, lowland chernozem with carbonate films
14, lowland chernozem with carbonate films, salty in deeper horizons
15. meadow chernozem
16, meadow chernozem salty in deeper horizons
17, alluvial chernozem
18, solonchak
19, solonchak-solonetz
20, meadow solonetz
21, meadow solonetz in transition to steppe soil
22, solonetz-like meadow soil
23, meadow soil
24, alluvial meadow soil
25, marshy meadow soil
26, lowmoor, fen soil
27, drained and cultivated fen soil
28, soils of swamp forests
29, recent alluvial soil

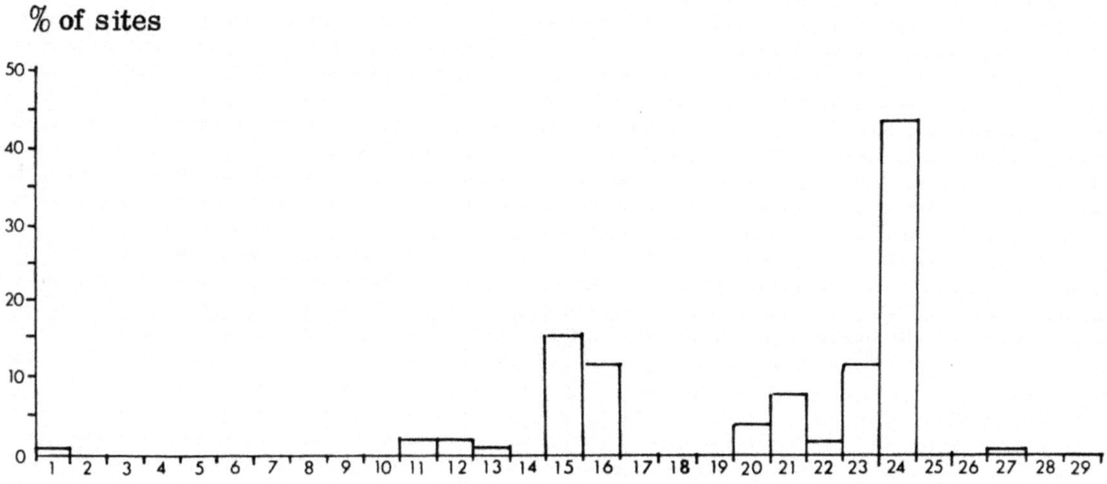

a) Körös culture

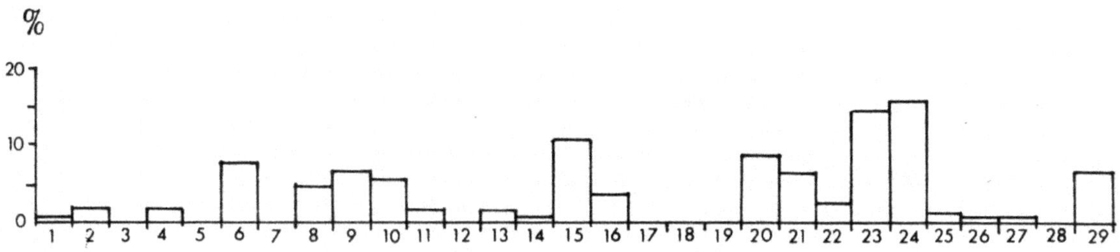

b) Alföld Linear Pottery culture

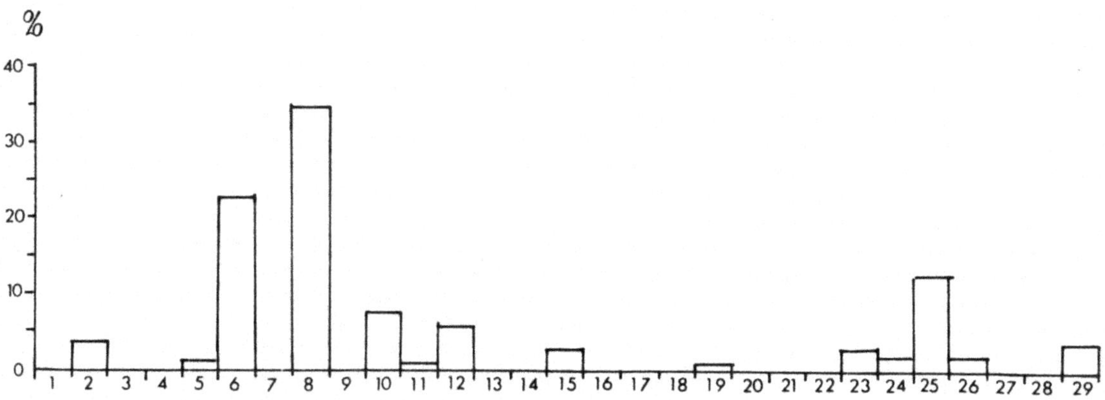

c) Transdanubian Linear Pottery culture

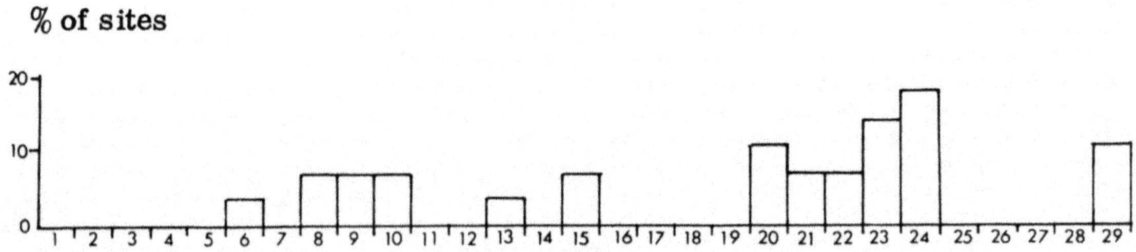

d) Early ALP

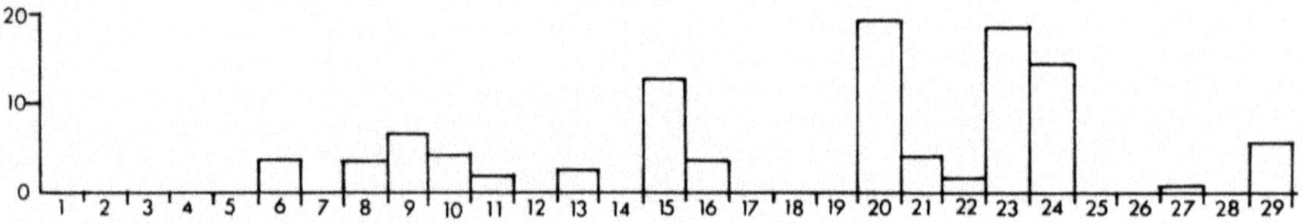

e) Middle ALP

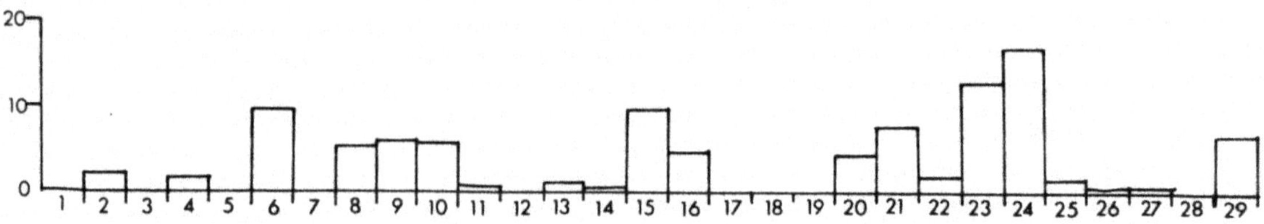

f) Late ALP

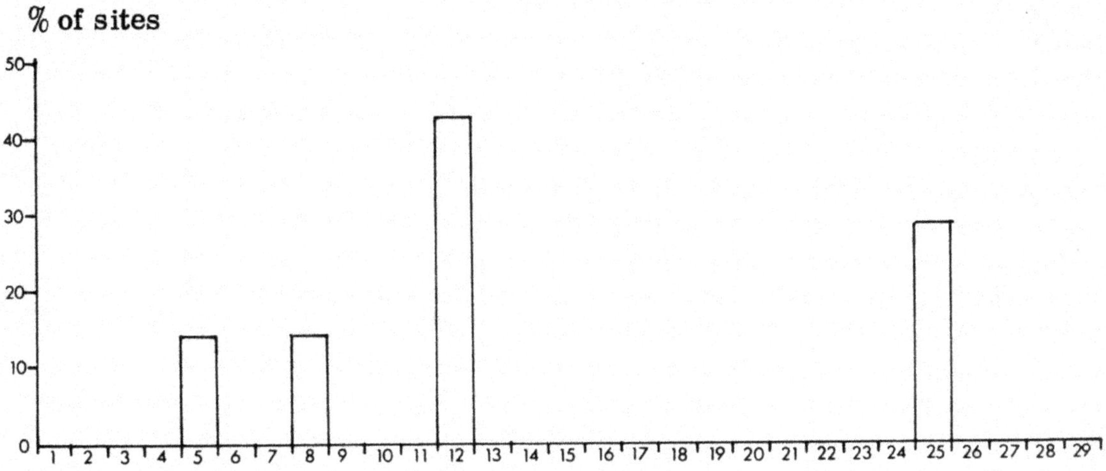

h) Early TLP

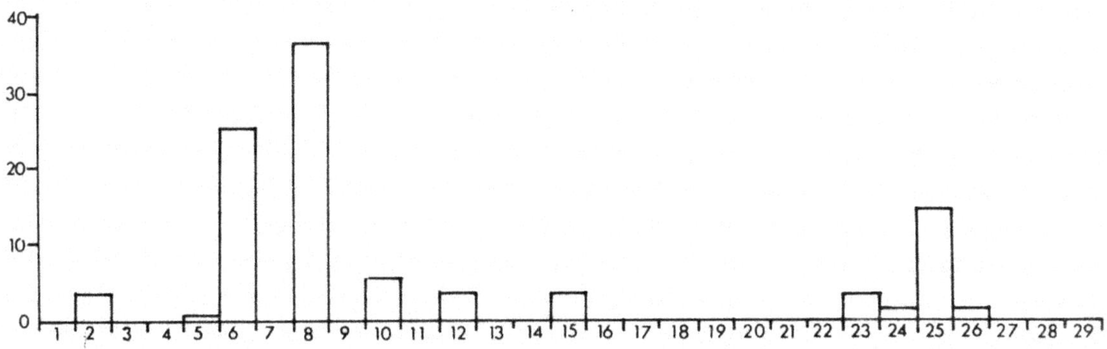

i) Middle TLP

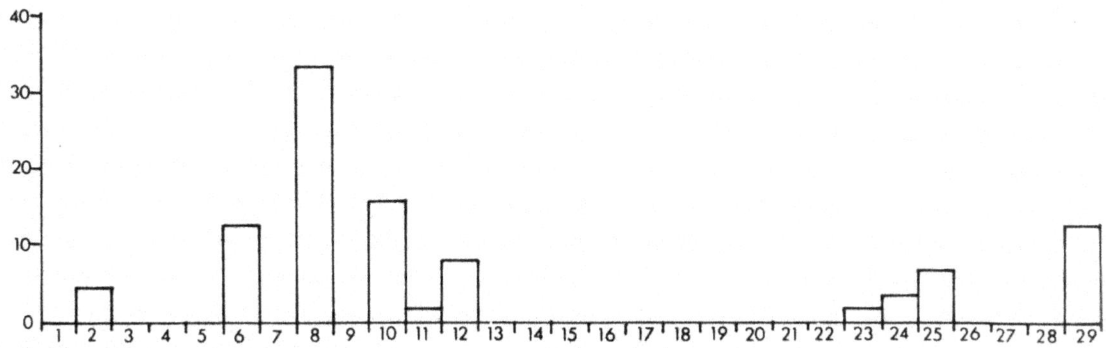

j) Late TLP

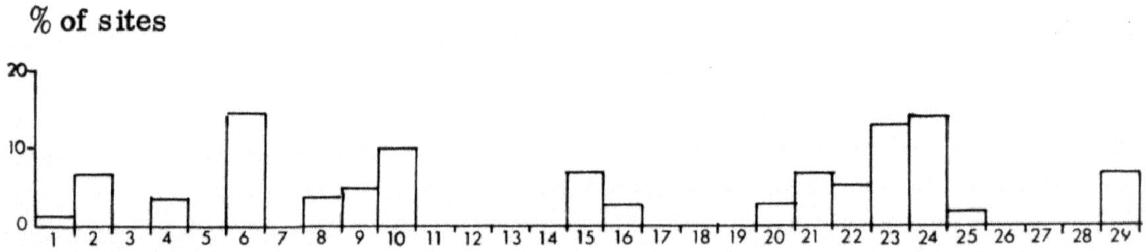

k) Late AVK

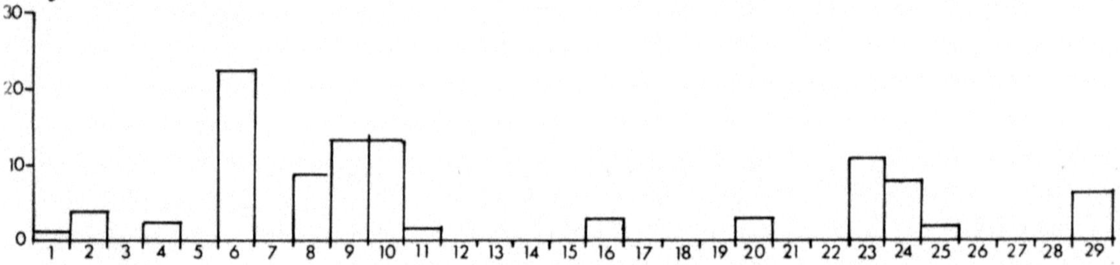

l) Tiszadob

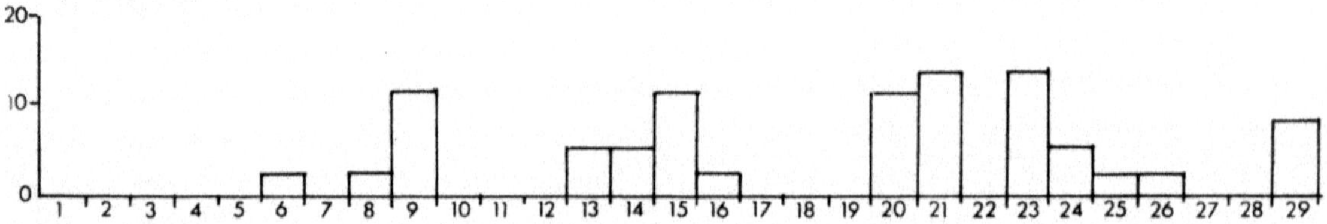

m) Esztár

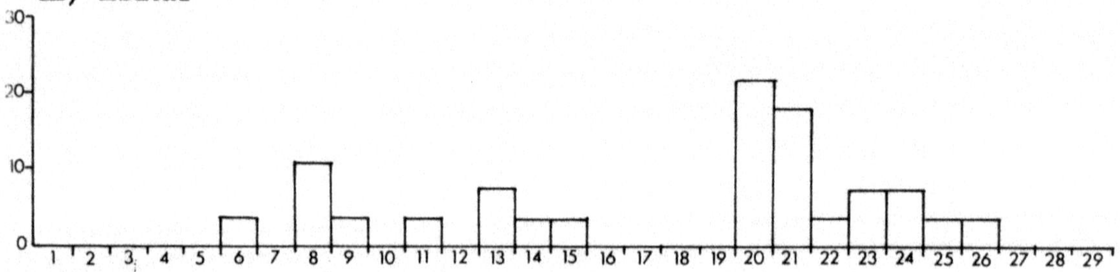

n) Szilmeg

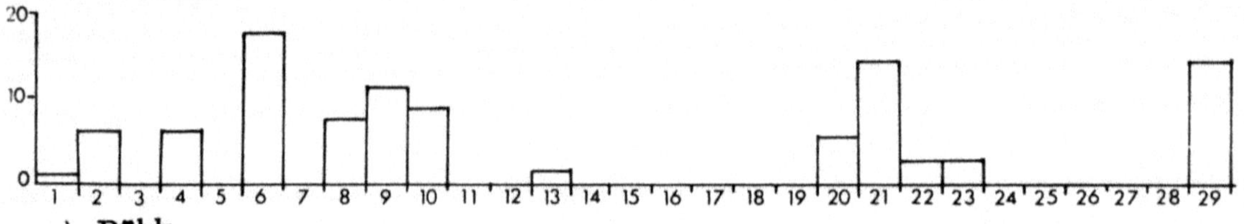

o) Bükk

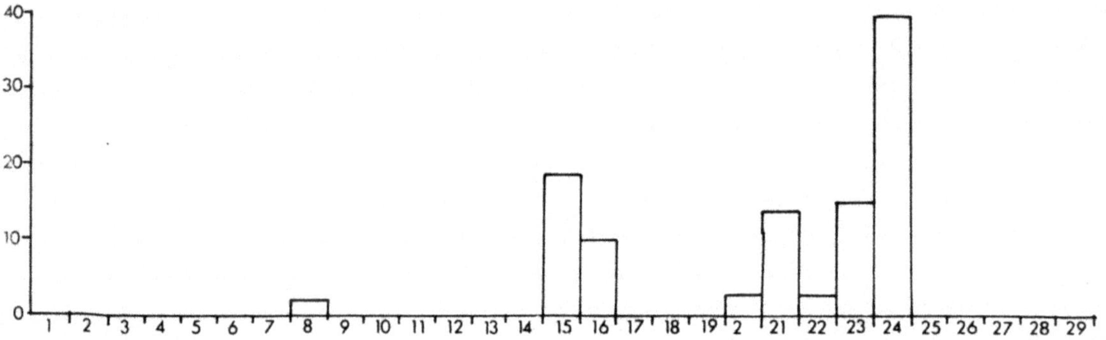

p) Szakálhát

Table 8 Percentage point difference between expected and observed distributions of settlements in relation to soils

Soil	Körös culture			ALP		TLP	
	Expected %	Observed %	Ppd	Observed %	Ppd	Observed %	Ppd
1	4.26	0.76		0.26			
2	2.36			1.56		3.30	-0.94
3	0.16						
4	0.30			1.30	-1.00		
5	0.19					0.95	-0.76
6	18.31			7.53		22.64	-4.33
7	1.90						
8	9.75			4.41		34.43	-24.68
9	2.61			6.23	-3.62		
10	5.41			5.19		7.08	-1.67
11	3.01	1.53		1.30		0.47	
12	7.50	1.53				5.66	
13	2.42	0.76		1.56			
14	0.90			0.52			
15	6.66	15.27	-8.61	10.65	-3.99	2.83	
16	3.29	11.45	-8.16	3.64	-0.35		
17	0.41						
18	0.16						
19	0.87					0.47	
20	2.50	3.82	-1.32	8.83	-6.33		
21	2.91	7.63	-4.72	6.23	-3.32		
22	1.82	1.53		2.08	-0.26		
23	6.57	11.45	-4.88	14.55	-7.98	2.83	
24	6.33	43.51	-37.18	15.85	-9.52	1.89	
25	3.10			1.04		12.26	-9.16
26	1.03			0.26		1.42	-0.39
27	0.90	0.76		0.26			
28	0.35						
29	4.02			6.75	-2.73	3.77	
Total Ppd			64.87		39.10		41.93

Table 8 (contd.)

Soils	ALP					
	Early		Middle		Late	
	Observed %	Ppd	Observed %	Ppd	Observed %	Ppd
1					0.40	
2					2.38	-0.02
4					1.98	-1.68
6	3.45		3.17		9.52	
8	6.90		3.17		5.16	
9	6.90	-4.29	6.35	-3.74	5.95	-3.34
10	6.90	-1.49	3.97		5.56	-0.15
11			1.59		0.79	
13	3.44	-1.02	2.38		1.19	
14					0.79	
15	6.90	-0.24	12.70	-6.04	9.52	-2.86
16			3.17		4.76	
20	10.34	-7.84	19.05	-16.55	4.37	-1.87
21	6.90	-3.99	3.97	-1.06	7.14	-4.23
22	6.90	-5.08	1.59		1.98	-0.16
23	13.79	-7.22	18.25	-11.68	12.70	-6.13
24	17.24	-10.91	14.29	-7.96	16.67	-10.34
25					1.59	
26					0.40	
27			0.79		0.40	
29	10.34	-6.32	5.56	-1.54	6.75	-2.73
Total Ppd		48.40		48.57		33.51
	TLP					
2			3.77	-1.41	4.76	-2.40
5	14.28	-14.09	0.63	-0.44		
6			25.16	-6.85	12.70	
8	14.28	-4.53	36.48	-26.73	33.33	-23.58
10			5.66	-0.25	15.87	-10.46
11					1.59	
12	42.86	-35.36	3.77		7.94	-0.44
15			3.77			
23			3.14		1.59	
24			1.26		3.17	
25	28.58	-25.48	14.47	-11.37	6.35	-3.25
26			1.89	-0.86		
29					12.70	-8.68
Total Ppd		79.46		47.91		48.81

(contd.)

Table 8 (contd.)

Soils	Late AVK		Late ALP Tiszadob		Esztár	
	Observed %	Ppd	Observed %	Ppd	Observed %	Ppd
1	1.24		1.49			
2	6.17	-3.81	4.48	-2.12		
4	3.70	-3.40	2.99	-2.69		
6	14.81		22.39	-4.08	2.78	
8	3.70		8.95		2.78	
9	4.94	-2.33	13.43	-10.82	11.11	-8.50
10	9.88	-4.47	13.43	-8.02		
11			1.49			
13					5.55	-3.13
14					5.55	-4.65
15	6.17				11.11	-4.45
16	2.47		2.99		2.78	
20	2.47		2.99	-0.49	11.11	-8.61
21	6.17	-3.26			13.89	-10.98
22	4.94	-3.12				
23	12.35	-5.78	10.45	-3.88	13.89	-7.32
24	13.58	-7.25	7.46	-1.13	5.56	
25	1.24		1.49		2.78	
26					2.78	-1.75
29	6.17	-2.15	5.97	-1.95	8.33	-4.31
Total Ppd		33.57		35.18		53.70
	Szilmeg		Bükk		Szakálhát	
1			1.28			
2			6.41	-4.05		
4			6.41	-6.11		
6	3.57		17.95			
8	10.72	-0.97	7.69		1.22	
9	3.57	-0.96	11.54	-8.93		
10			8.97	-3.56		
11	3.57	-0.56				
13	7.14	-4.72	1.28			
14	3.57	-2.67				
15	3.57				18.29	-11.63
16					9.76	-6.46
20	21.44	-18.94	5.13	-2.63	2.44	
21		-14.95			13.41	-10.55
22	3.57	-1.75			1.22	
23	7.14	-0.57	14.10	-7.53	14.63	-8.06
24	7.14	-0.81	2.57		39.03	-32.70
25	3.57	-0.47	2.57			
26	3.57	-2.54				
29			14.10	-10.08		
Total Ppd		50.72		42.89		69.41

hydromorphic soils and chernozems, and in some classifications are actually regarded as hydromorphic soils (Szabolcs, 1961).

The ALP settlements are found on a much greater variety of soil types. Nevertheless, the largest number of settlements are on the same soil types as the Körös sites: 23% on alluvial soils; 17% on alkali soils; and 16% on hydromorphic soils. 14% are on meadow chernozems. In contrast to the Körös culture, however, 25% of the ALP settlements are found on forest soils. The remaining small percentage of sites are distributed as follows: 2% on chernozems, 2% on rendzinas and less than 1% on windblown sand

The majority (65%) of TLP sites are found on forest soils. The soil type which has the highest percentage of settlements within this group is Ramann's brown forest soil (34%). As to the rest of the sites: 17% are on hydromorphic soils, 3% on meadow chernozems, 6% on chernozems and 3% on rendzinas.

The distribution of settlements in the Early, Middle and Late phases of the ALP shows little deviation from the overall pattern described for the ALP; the highest percentage of sites in all three phases are found on alluvial, hydromorphic and alkali soils. Forest soils are occupied from the Earliest phase onwards. The number of soil types occupied by the settlements is greatest in the Late phase.

In the Early TLP, the highest percentage of sites are on chernozem soils. However, the number of settlements in this phase is very small, and the pattern is probably somewhat distorted. In both the Middle and Late phases,the majority of sites are found on forest soils; the soil type most preferred for settlement is Ramann's brown earth.

Of the late groups of the ALP, the settlement pattern of the Szakálhát group is the most distinct; similarities to the Körös distribution are very noticeable. The distribution patterns of the Late AVK, Tiszadob and Bükk groups are similar and show a definite contrast to the Esztár and Szilmeg patterns.

Chi-square tests suggest that the probability of chance distribution in either culture is less than 0.05%. Percentage point differences between the expected and observed patterns of distribution are shown in Table 8. Most of the values are fairly high: they range from 33.51 to 69.41. The lowest value (33.51) is obtained for the Late ALP. When the regional groups are analysed separately, however, the figures are higher.

Comparing the distributions of the settlements of the three cultures, the results show clearly contrasting patterns for the Körös and TLP cultures, with the ALP occupying a transitional position between the other two. As we have seen the majority of TLP sites are on forest soils, while the majority of Körös sites are found on alluvial and hydromorphic soils. In the ALP, sites appear on forest soils from the earliest phase on, but with the exception of the Late Tiszadob and Bükk groups these settlements never constitute the majority.

The almost exclusive occupation of alluvial-hydromorphic soils in the Körös culture is a reflection of the location of the settlements in or near the floodplains. As mentioned before, the Körös sites are located in climatically dry areas, but in regions where there is an ample supply of surface and subsurface moisture. Almost all the soil types occupied by the Körös settlements show evidence of having been formed under the effect of at least periodically excessive moisture.

In the TLP culture, where settlements are mostly located in higher rainfall areas, the importance of the alluvial and hydromorphic soils is much less: only about 25% are located on these soils. It should be pointed out, however, that although the majority of the settlements are on forest soils, there is always a certain percentage of alluvial or hydromorphic soil in the vicinity of the sites. As the Linear Pottery people practiced a mixed economy, the inclusion of several soil types in the territory of the settlements is hardly surprising. Possible land use and management practices will be discussed in the next chapter.

As yet, the relationship of Linear Pottery settlements to soil type has been analysed in very few areas outside Hungary. One such area, however, is Poland. Kruk (1972-73) gives the following figures for the Danubian occupation of the Little Poland Highland region: 40% of the settlements are on muds and other soils on river alluvium; 35-38% on brown forest soils; 15% on chernozems; and 8-10% on rendzinas. The figures on the whole are similar to those obtained for the Linear Pottery cultures in Hungary. In the TLP alluvial and hydromorphic soils are of somewhat less importance and brown forest soils of more. The ALP figures differ more widely; the occupation of alluvial and hydromorphic soils is higher and that of forest soils is much lower.

The attraction of chernozem soils seems to have been even smaller in Hungary than in Poland. These soils, and meadow chernozems are excepted, were largely avoided not only by the Linear Pottery settlers, but also by those of the Körös culture. When settlements do occur on chernozems, they are always at the edge of the chernozem zone, near ot at the junction with another soil type.

As to the possible utilisation of the selected soils by the three cultures in Hungary, some suggestions can be made on the basis of the inherent characteristics of the soil types. The preference of the TLP settlers for Ramann's brown earth and,secondarily, for the lessivated brown forest soils would indicate that agricultural considerations were a major factor in the selection of the sites. Ramann's brown earth is the most fertile forest soil in Hungary, and the lessivated brown forest soil, though less fertile than Ramann's, is more productive than any other forest soil. In the case of either soil, a periodical replenishment of soil nutrients is required, however, for continuous agricultural production with high yields.

In the ALP, the second most frequently selected forest soil after the lessivated brown forest is the kovárvány forest soil. Kovárvány soils are exceptionally fertile for sandy soils. Because they are sandy they are also easy to work. Continuous production with good yields , as in the case of the other forest soils, requires the use of fertilizers. The podzolic, the strongly acid non-podzolic brown forest soils and the rendzinas are of limited value for agriculture. Their occupation, also limited, may have been connected with seasonal pasturing or with some other need. The rendzinas would have been especially favoured by sheep and goats.

Fig. 12. Distribution of TLP settlements in relation to soil types in the Marcal region. (Soil map after Góczán, 1971.)

- stony skeletal soil
- slightly humic sand
- rendzina
- black and red erubase soil
- lessivated brown forest soil
- lessivated brown forest soil with 'kovárvány'
- 'kovárvány' brown forest soil
- braunerden
- rust brown forest soil
- chernozem-brown forest soil
- meadow chernozem
- chernozem-type sand
- alluvial chernozem
- chernozem-meadow soil
- meadow soil
- alluvial meadow soil
- floodplain forest soil
- marshy meadow soil
- muck soil
- peat
- peaty muck
- chernozem-type gravelly alluvium
- colluvium of brown forest soils

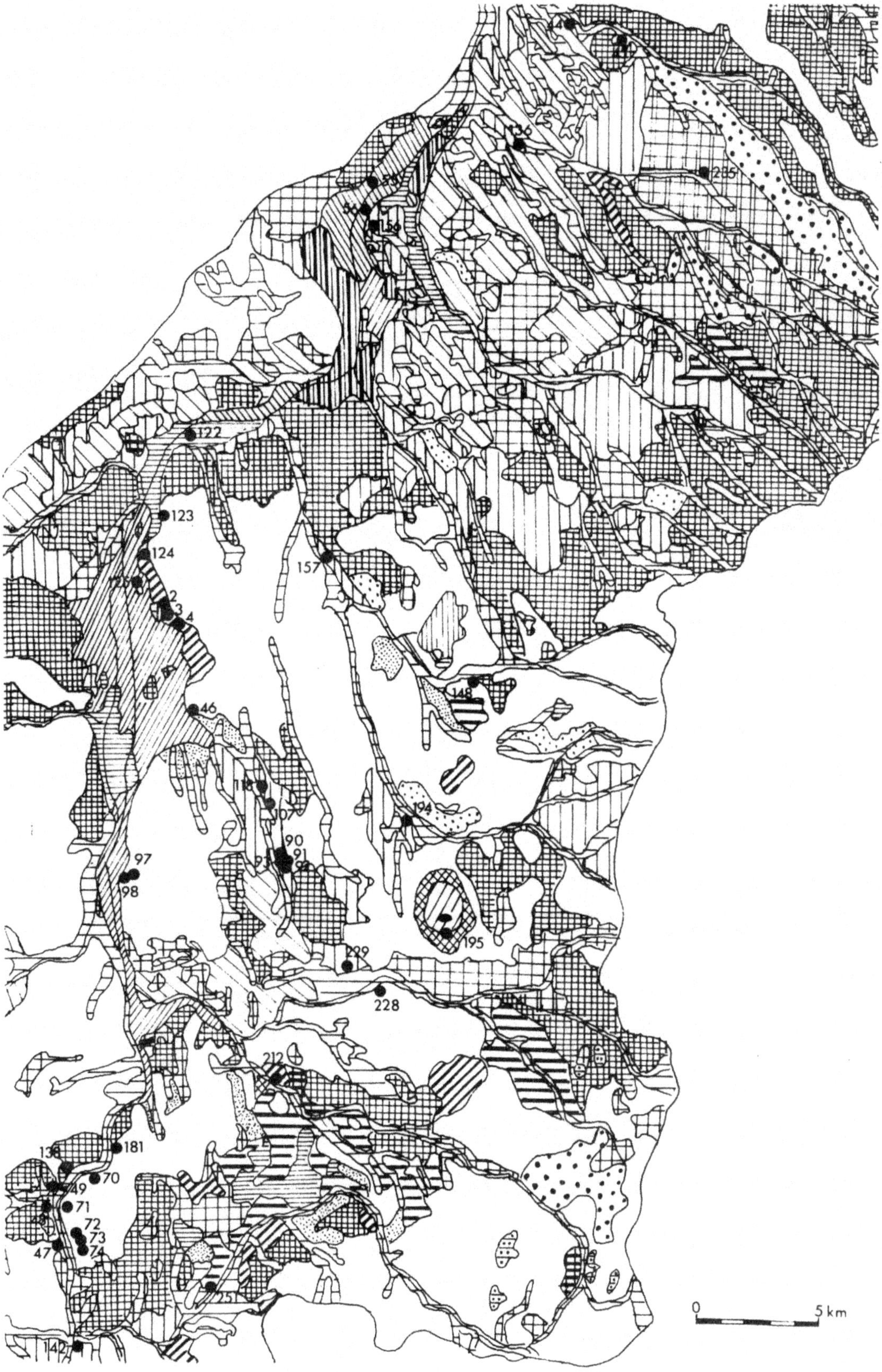

44
136
58
159
122
123
124
2
3
4
157
148
46
107
194
90
91
97
98
195
228
212
181
70
49
71
72
73
74
47
142
0
5 km

The alluvial and hydromorphic soils preferred by the Körös culture and to some extent also by the ALP are of great potential fertility. Although not strictly a hydromorphic soil, the meadow chernozem is the most fertile soil type in Hungary. In the case of the alluvial soils, the periodical inundation of the rivers in the Alföld is a good supplier of soil nutrients. Soil fertility can, however, vary a great deal even within a small area in the floodplain. Microtopographical differences in so far as they effect drainage, are important. To a certain extent soil texture is also connected with microtopography. Some meadow soils, especially those in low lying areas, are too heavy for easy cultivation. Impeded drainage and salt accumulation are also frequent obstacles to agricultural utilisation. The threat of flood disasters as a deterrent has already been mentioned. Good agricultural land in the zone of alluvial and meadow soils is fragmentary and also small in extent. Large scale agriculture has only become feasible in this region since the regulation of the rivers.

Although the soil map used in the analysis provides a general picture of settlement distributions, its relatively small scale does not allow the recognition of finer details. To supplement the information of the small scale map, the distribution of settlements has also been examined using large scale soil maps in two selected regions of Hungary. The two regions are: the Marcal basin in Transdanubia and the Tiszazug in the Alföld. The former is in the distribution area of the TLP, and the latter has settlements of both the Koros and ALP culture. These two areas have been selected because they both have been systematically surveyed by archaeologists and also because the availability of relatively detailed soil maps.

Settlement Distribution in the Marcal Basin

The Marcal basin is situated at the southeastern margin of the Little Plain, in the foothills of the Bakony Mountains. The intra-valley ridges running down from the mountains are covered by gravels and sands of old alluvial fans. The major landforms are the terraced alluvial fans east of the Marcal, the floodplains of the Marcal and the basalt capped monadnocks of the Somló and Ság.

The detailed soil map of the area has been prepared by Goczán (1971). In general, lessivated brown forest soils cover the higher terraced plains, braunerden the lower lying basins and slopes and alluvial and hydromorphic soils the river valleys and lower lying depressions. The distribution of settlements is shown in Fig. 12. The number of TLP sites in the area is 42. Table 9 shows the number and percentage of sites associated with each soil type.

On the soil map of Hungary, most of the sites in this area occur on lessivated brown forest soils, the only forest soil shown on the map. On Góczán's detailed map, on the other hand, only 28.57% of the settlements are associated with this soil type; the majority (54.76%) of the sites are found on variants of Ramann's braunerden. Although in the general distribution of TLP settlements the preferential association of the sites with braunerden is also evident, the example of the Marcal region suggests that the magnitude of the association is probably somewhat underestimated.

Of the 23 sites found on variants of Ramann's brown forest soils in the Marcal region, the largest number of sites occur on chernozem-brown forest

Table 9. Distribution of TLP settlements in the Marcal region according to soil type

Soil type	Number of sites	% of sites
lessivated brown forest	11	26.19
lessivated brown forest with kovárvány	1	2.38
brown forest	7	16.67
chernozem-brown forest	11	26.19
colluvium of brown forest	5	11.90
erubase	1	2.38
meadow-chernozem	1	2.38
chernozem-meadow	1	2.38
meadow	1	2.38
marshy meadow	1	2.38
muck	2	4.76
	42	100.00

intergrades. The genesis of this soil type, as previously mentioned, is still imperfectly understood. When the first detailed land use map of the Marcal region was drawn up in 1782-84, the present chernozem-brown forest soil areas were already without forest cover. An analysis of the settlement history of the region would probably show that cultivation in these areas had started long before the eighteenth century. Góczán is inclined to see human influence as the major factor in the transformation of these soils in the Marcal region.

While it is unlikely that the transformation could already have occurred during the lifetime of the TLP, it is of interest that the TLP sites are, almost without exception in those areas which are shown as cultivated fields on the eighteenth century map. At this time the extent of cultivated fields in the Marcal region is still very small; most of the region is covered by oak forests. The occurrence of TLP settlements almost exclusively within the area of distribution of the eighteenth century fields again suggests the importance of agricultural considerations in the selection of the sites. Unfortunately, the composition of plant remains from the sites is not yet known.

11 of the sites are associated with lessivated brown forest soils. In those areas where the settlements are found the soil is usually fine textured and contains less gravel than in other areas. The deeper soil profile ensures a better supply of nutrients and a more favourable water economy for plant growth. These soils are also easier to cultivate and give higher and more consistent yields than their gravelly variants.

In summary, the settlement pattern observed in the Marcal region shows little deviation from the general pattern of TLP settlements in Hungary. The overwhelming majority of the sites are located on forest soils but always in the proximity of alluvial or hydromorphic soils. It is usually the most fertile soil type which is selected for the location of the settlements. In the Marcal region where rainfall is relatively high, the settlements generally occur in drier micro-areas, that is, in lower lying plains or on gentle slopes where conditions are favourable for the development of Ramann's braunerden. In the Marcal basin, real loess is limited in extent. The settlements are located on loess-like sediments, usually on finer textured calcareous parent materials. Shallow gravelly soils are avoided.

Settlement Distribution in the Tiszazug Region

The Tiszazug region is situated at the confluence of the Tisza and Körös rivers. The Tisza borders the area in the west and the north, and the Körös in the south. The present floodplains of the two rivers are bordered by the remnants of two Holocene terraces above which rises the loess plateau of the Nagykunság. The loess plateau is capped in places by smaller blown sand areas. Both the upper and lower terraces are dissected by ridges and depressions and are surrounded by cutoff oxbows. The settlements of the Körös and ALP cultures are usually located on the terraces. Many of the present day villages are also found on the terraces.

The detailed soil map of the area was compiled by Nagy (1954). In general, the floodplains and oxbow channels are covered by alluvial and meadow soils, the terraces by alkaline soils and the loess surfaces by meadow chernozems.

The distribution of settlements is shown in Fig. 13. The number of Körös sites is 22, and the number of ALP sites in the area is 36. 29 of the 36 ALP settlements belong to the Szakálhát group. The number and percentages of Körös sites associated with each soil type is shown in Table 10, and those of the Szakálhát group are given in Table 11.

On the detailed map of the Tiszazug region, 63.64% of the Körös sites are located on alkaline soils, and the remaining 36.36% are found on meadow-chernozems. In contrast, 65.51% of the Szakálhát settlements are on meadow chernozems, and 34.49% on alkaline soils. The number of earlier ALP sites is too small to give a reliable estimate; 6 of the 7 sites are, however, also on meadow chernozems.

Meadow chernozems, as we have seen, are the most fertile soil types not only in the area but in the whole of Hungary. Because the capillary zone of the ground water table is usually within the reach of the plant roots, high yields are assured on these soils even in periods of drought. The productivity of the alkaline soils depends on several factors. The most fertile soils in the group are the chernozems with alkali subsoils and the deep solonetzes. Most of the Körös and Szakálhát sites located on alkali soils are on deep solonetzes.

The depth of the salt free upper horizon, and thus soil fertility is to a large extent dependent on the microtopographical position of the soil. According to Nagy's observation in the Tiszazug region, even very slight differences in elevation, and sometimes the difference is no more than a few centimeters, can be decisive. Even when the salt concentration is the same, crops tend to suffer more in the lower lying areas. Szabolcs (1961) attributes this to the deterioration of the physical properties of the soil which is brought about by frequent inundations.

The position of the Körös and ALP sites on former levees would have been a definite advantage: the higher and coarser textured levees, even when alkaline, would bring better yields than the surrounding lower lying areas. The low salt content and the presence of good quality land on the levees at least in the Tiszazug region is indicated by present-day land use. Almost all the sites are under maize today (pers. observation). Maize is not very tolerant of alkali conditions and cannot be grown even on medium solonetz soils in Hungary (Szabolcs, 1966). The maize fields which cover the sites are located on the ridge tops and slopes and terminate abruptly when low lying ground is reached. In consequence, the size of the fields depends on the size of the ridges and is often very small.

Szűcs (1967) in his study of chernozem soils in the Nagykunság mentions the regular occurrence of meadow chernozem soils on the banks of old river beds and oxbows. The chernozems follow the river in a narrow strip: their width across is no more than 100 to 200m. Beyond this narrow belt, alkali soils usually are the dominant soil type.

In some areas of the Alföld, however, more alkaline levees have also been observed. In these, the alkalinisation is probably secondary and has occurred since the time of the Körös and ALP cultures. At Szamossályi, for example, the salty layer lies above the cultural horizon, postdating the ALP site (Stefánovits, 1953). Kalicz has made similar observations at other Körös and ALP sites near Hódmezővásárhely and Tiszavasvári (pers. comm.).

Although the levees were probably salt free at the time of the Körös and Linear Pottery occupation, the surrounding low-lying land would have been, if drainage was impeded, alkaline. Alkaline soils were present in the Alföld before the regulation of the rivers, although they were probably less extensive (Szabolcs, 1961).

The analysis of site distribution suggests that soil alkalinity would have been especially problematic in the case of the Körös sites, the majority of which are located in areas of salt affected soils. While the levee tops would have been suitable for crop production, the surrounding area would have been of limited fertility even when not under water. The emphasis on sheep breeding, which is evident from the bone remains, could have been a response to the prevalence of alkali soils in the neighbourhood of the settlements. Good quality agricultural land due to the nature of the terrain would have been limited in extent. The best areas for the location of the fields would have been on the levee tops and slopes with their fertile soils and favourable drainage.

The position of the Szakálhát settlements is similar to that of the Körös sites. There is, however, a slight shift in emphasis as regards to soil type. The majority of the Szakálhát sites are located in areas with access to larger tracts of salt free soils. The change is small, as most of the sites are still located on river banks, although in one instance the site is located further away from the river (Kalicz, 1957). The greater occupation of meadow chernozems may have been connected with expanding agricultural production or with some changes in herd composition. Unfortunately, none of the Szakálhát settlements has been excavated yet, and the composition of plant and animal remains is still unknown.

In summary, according to the detailed analysis, more Körös sites seem to be associated with alkali soils and more Szakálhát settlements with meadow chernozems than may have been expected on the basis of the small scale map.

Fig. 13. Distribution of Kőrös and ALP settlements in relation to soil types in the Tiszazug region. (Soil map by Nagy, 1954.)

Non calcareous sand

Calcareous sand

Non calcareous loam

Calcareous loam

Non calcareous clay loam

Complex areas

hh: humus sand

kh: /meadow/chernozem - with medium humus layer

lh: coarse sand

vh: /meadow/chernozem - with shallow humus layer

hő: humous alluvial soil

fő: recent alluvial soil

rt: meadow soil

ksz: deep solonetz

msz: solonetz

szm: chernozem, solonetz in deeper layers

szr: meadow soil, solonetz in deeper layers

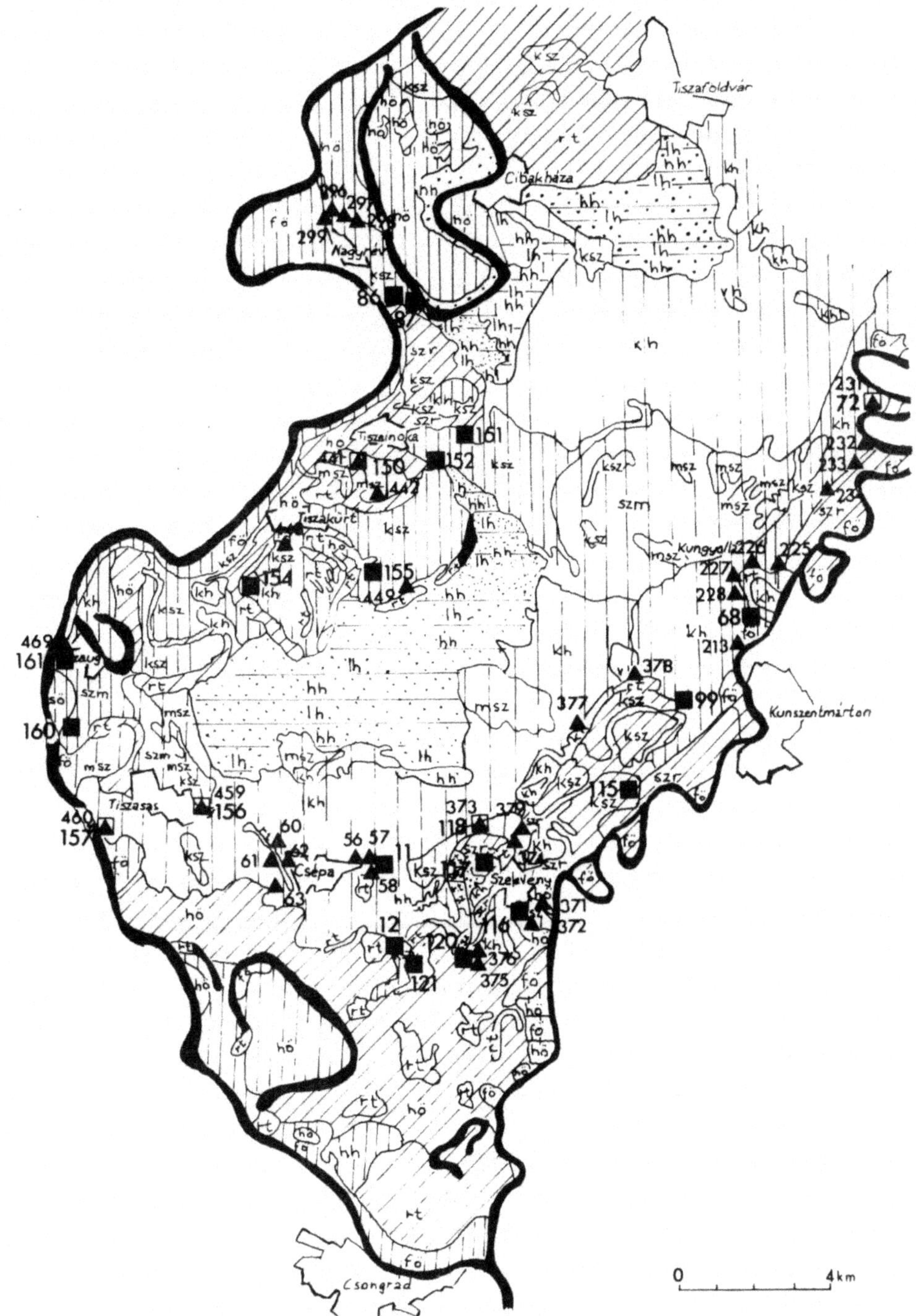

■ Körös settlements

▲ ALP settlements

Table 10 Distribution of Körös settlements in the Tiszazug region according to soil type

Soil type	Number of sites	% of sites
meadow chernozems (on calcareous loam)	3	13.63
meadow chernozems (non-calcareous loam)	4	18.18
meadow chernozems (clay loam)	1	4.55
chernozems, alkaline in deeper layers	2	9.09
deep solonetz	11	50.00
solonetz	1	4.55
	22	100.00

Table 11 Distribution of Szakálhát settlements in the Tiszazug region according to soil type

Soil type	Number of sites	% of sites
meadow chernozems (on calcareous loam)	10	34.48
meadow chernozems (non-calcareous)	9	31.03
chernozems, alkaline in deeper layers	2	6.90
deep solonetz	7	24.14
solonetz	1	3.45
	29	100.00

The analysis also suggests that soil relationships in the belt of the alluvial and meadow soils are far more complex than the small scale map of the country would indicate. With the exception of the loess uplands, which are uniformly covered by chernozems, the number of soil changes even within a small area in the Alföld is very great. In consequence good agricultural land is greatly circumscribed. Microtopography in this area is often a useful guide to soil conditions.

Soil Changes since the Atlantic Period

As the analysis of settlement locations was based on the present-day soil map of Hungary, the evaluation of the results would be incomplete without some consideration of the extent and nature of the soil changes that may have occurred since the Atlantic period.

Since soils are perceived as the product of a number of environmental factors, changes in these factors may be expected to produce corresponding changes in soil formation. In addition, from the Atlantic period onward, man himself has to be considered as a potential soil-forming agent. In the case of some of the soil types in Hungary, there is general agreement as to their age and development, while in the case of others, these issues are still matters of controversy.

It is generally agreed that the brown forest soils have been the climax soil types of the moister uplands and hilly regions since post-Pleistocene times. According to Hungarian pedologists, the oldest is the lessivated brown forest soil, which had developed on older loess and periglacial gravel surfaces during the Preboreal (Góczán, 1971). Braunerden are generally found on younger surfaces and are considered to date from the beginning of the Atlantic period (Góczán, 1972).

The rendzinas, like the brown forest soils, are post-Pleistocene in age. According to Stefánovits (1971), Mediterranean-type of soil formation has played an important role in the genesis of the soil mantle of Hungary, especially in the more submediterranean southwest. It should be emphasised, however, that the rendzinas are intrazonal in Hungary, and parent material was probably more important in their formation than climate.

The soil types whose age and development are still controversial are the chernozem and alkali soils. In both cases the question of age is closely related to the problem of defining the extent of anthropogenic influence on soil formation.

Gradmann (1906; 1933) was one of the first to observe a connection between the present distribution of steppe plants and chernozem soils and neolithic and bronze age settlements in Central Europe. He suggested that the connection was due to the preference of prehistoric man to settle in areas where the primeval forest had already been destroyed by the warmer and drier climate of the Subboreal.

Gradmann's theory has been refuted on several grounds. The establishment of the first agricultural settlements took place in the Atlantic period when according to pollen investigations most of the central European loess lands were covered by mixed oak forests (Garnett, 1945). Even in the Subboreal many of these areas were still carrying a forest vegetation. As regards eastern Europe, Grichuk was able to show that a Subboreal xerothermic period had never existed there, but that the last steppe communities in the recent forest belt of eastern Europe were already displaced by the early postglacial immigration of forests and since then never succeeded in coming back spontaneously (Frenzel, 1966). The rejection of Gradmann's theory does not, however, solve the problem of the origin of European chernozem soils. The evidence merely suggests that the climatic change of the Subboreal cannot have been the main agent in their formation.

Kruk (1971), in his study of the loess area of the Nida basin, Poland, assigns the major role to man. In his view, the economic practices of the TRB and Corded Ware peoples resulted in large scale deforestation and thus led to the initiation of chernozem formation. In direct opposition to Kruk's is Wilhelmy's (1950) view, according to which the chernozem soils of central and eastern Europe are relics of the Late-Glacial and Post-Glacial periods and predate the appearance of the first agricultural communities. Examining the distribution of the Linear Pottery settlements in central Germany and the distribution of neolithic and copper age sites in Bulgaria, he finds that the settlements are almost always found on degraded chernozems, but only very rarely on real chernozems. As the tough sod-cover of the chernozems is very difficult to break up with primitive agricultural implements, the lack of settlements in these areas would be consistent with the view that some chernozem areas had already existed at the time of the earliest farming settlements. The Stichbandkeramik date of some chernozem soils found in dwelling pits in Germany would seem to strengthen Wilhelmy's argument. Lózek (1967), reviewing the evidence from Czechoslovakia, comes essentially to the same conclusion: chernozems were present in the driest enclaves before the Subboreal.

Hungarian paleobotanists and pedologists consider the present-day chernozems to have originated in the Boreal phase. 'It was in this period that the climate and vegetation offered the most favourable conditions for the formation of climazonal chernozems in the lowlands of Hungary (Somogyi, 1964, 38-39). The fact that the leached or degraded chernozems are restricted in their distribution to the margins of the Great Plain (Stefánovits, 1971) suggests that the chernozem soils in the driest areas of the Alföld and Mezőföld were not invaded by forest vegetation during later climatic phases. Further evidence comes from buried chernozem soils which have been found under several Early Bronze age khurgans in the Alföld (Borsy, 1966-67). The burial in one of these tumuli has been dated to 2370 bc (Ecsedy, pers. comm.). As, in Borsy's opinion, the full development of these soils had required something like 2000 to 3000 years, a pre-Atlantic date for their formation seems likely. The presence of chernozem soils during the Atlantic is indirectly confirmed by the virtual lack of settlements on the high loess plateaus. As Loźek has suggested man's influence probably contributed to the continued existence of the early chernozem-steppes, but climatic factors at least in eastern Europe seem to have been more important in their formation.

Man's influence on the formation of alkali soils has been the subject of debate in Hungary since the regulation of the rivers. Although the precise nature of the changes that occurred after the alteration of drainage conditions is still disputed, it is generally agreed that salt affected soils had already existed before the regulations. It has been observed that the distribution of solonetz soils is closely connected with those areas which had been temporarily inundated prior to the regulations (Szabolcs, 1961). According to Máté (1955), these soils had already possessed a solonetz profile in preregulation times and did not evolve from solonchaks by leaching. Some leaching did take place even so with the general sinking of the ground water table, but soil development mostly depended on local conditions. In areas of good drainage soil formation led to steppe like soils which preserved their alkalinity only in the deeper horizons. Where, on the other hand, drainage was impeded by an impervious layer near the surface solonchak-solonetzes developed. In the Hortobágy, for instance, large scale irrigation has resulted in the formation of secondary solonchaks over large areas.

Although the regulations undoubtedly added to the extent of alkali soils in the Alföld, not all salt affected soils were created by them: in the opinion of pedologists and paleobotanists alkali soils in Hungary have existed since the Boreal period. According to Soó (1959), most of the solonchak soils of the Alföld are descendants of the solonchak soils that were formed during the steppe climate of the Boreal. Before the Subboreal downcutting of the rivers, the areal extent of those soils which were annually inundated or were under the influence of the fluctuations of a high ground water table was much greater. Solonchak soils occupied the drainless depressions, and all those areas where the ground water table is above 5 or 6 m today were covered by solonetz soils during the Boreal period (Somogyi, 1965).

In the Atlantic phase, the moister climate presumably led to a further elevation in the level of the ground water table. Where drainage was no longer impeded leaching or partial leaching resulted in the transformation of some solonchaks into solonetzes. According to Nagy (1954), this was the period when some of the deep solonetzes of the Tiszazug region were formed. While some solonetzes were entirely freed from their salt content during this period, the possibility for the formation of new solonetz soils also existed due to the elevation of the ground water table (Somogyi, 1965).

The Subboreal phase brought a number of changes in the development of alkali soils. With increased precipitation and the downcutting of the rivers, alkali soils disappeared entirely in some areas. In other areas, however, especially near the newly cut off oxbows, alkali soils began to develop. The changes that occurred in the vegetation cover of these areas due to increased salt concentration in the soil have already been mentioned. The formation of alkali soils on some of the archaeological sites may have taken place in this or the subsequent Subatlantic phase.

One soil type whose extent was probably much smaller in the Atlantic period than it is today is the meadow chernozem. Soil investigations have shown that most of these soils were still meadow soils before the regulation of the rivers (Szabolcs, 1961). But even before the sinking of the ground water table had transformed them into meadow chernozems, these soils differed in many respects from today's meadow clays. They were loamy rather than clayey in texture; had better physical properties and greater potential fertility. Because they were lighter, they were easier to work. It is probable that these so-called meadow-loams covered many of the levees which were occupied by the Körös and ALP settlements (Várallyay, pers. comm.).

Man's direct or indirect influence on the formation of chernozem and alkali soils has already been mentioned. In addition, man's agricultural activity was probably responsible for the formation of some of the chernozem-brown forest soil intergrades. On some dolomite and limestone slopes grazing has probably been a factor in the accelerated development of the rendzinas (Góczán, 1971). It is also possible that long term levee occupation has contributed to the transformation of some meadow soils into meadow chernozems well before the regulation of the rivers. All these changes, however, suggest long term interference, and although some of them may have been initiated in the Atlantic period, it seems unlikely that the effect of Körös and Linear Pottery occupation would have resulted in drastic soil changes. The effect was probably local and small scale in contrast to later developments.

RECONSTRUCTION OF THE ECONOMIC BASE

In the previous section, the settlement distributions of the Körös, ALP and TLP cultures have been analysed in relation to the 'natural' environment. In this chapter, the observed patterns of site location are used in an attempt to reconstruct the subsistence and land use patterns of the three cultures. The reconstruction is essentially an interpretation of the environmental potentialities and limitations of the immediate vicinity of the sites together with an evaluation of the archaeological remains, especially of food remains, found on the sites.

Körös Culture

The settlements of the Körös culture are located in the Alföld, which climatically is the hottest and driest region in Hungary. The zonal vegetation in the driest parts is the steppe or forest steppe, and the zonal soil type is the chernozem. While 87%, that is the overwhelming majority of the Körös sites, are in areas where humidity is insufficient over the entire year, less than 4% of the sites are found in the chernozem steppes. The majority of the settlements are located in azonal situations, mostly in or at the edge of the floodplain.

It has been calculated that on the average 70% of the land within a radius of 2.5 km of the sites falls within the floodplain. Even in the remaining 30% of the area water seems to have been an important factor in shaping the soil and vegetation cover; 95% of the Körös sites are located on soil types where water has been directly or indirectly an agent of soil formation. The soil map in this case is a sensitive indicator of past environmental conditions. The 95% figure includes not only the present floodplain, but also those areas which had a high ground water table before the regulations of the rivers (meadow chernozems) and which had originally, in the view of the geomorphologists, belonged to the floodplain proper before the Subboreal downcutting of the rivers.

The sites are located immediately by the rivers, usually on levees or dunes. The settlements are generally very long and narrow. In some cases the length of the site is 300 to 400 m, while its width does not exceed 30 or 40 meters (Kalicz, 1957). Even more extreme in this respect is the Körös site at Dévaványa-Katonaföldek, where the settlement is some 800 m long but only 50-100 m wide (Ecsedy, 1972). The location of the sites in a narrow bend along the river is a characteristic feature of the Körös culture and is probably connected at least in part with the importance of the river in the life of the Körös people.

The composition of food remains from the sites shows a large degree of reliance on the resources of the river and floodplain. Of the three earliest farming cultures in Hungary, the use of aquatic resources seems to have been the greatest in the Körös culture. One way of exploiting the riverine and floodplain environment was by fishing and the gathering of freshwater shellfish, water turtle and snails.

Where data are available, the percentage of fishbones in the total animal bone material is fairly high. At Maroslele-Pana it is 12.6% (Bökönyi, 1964), at Gyálarét-Szilágyi major 14.5 (Bökönyi, 1974), at Deszk-Olajkút 16.7% (Trogmayer, 1968), and at Röszke-Ludvár 19.26% (Bökönyi, 1974). Catfish (Silurus glanis L.) and pike (Esox lucius) are the only two species which could be identified so far. Very large catfish bones are not infrequent,and according to Bökönyi some individuals must have weighed near 200 kg.

Besides fishbones large quantities of fishscales have also been found at the sites. At Röszke-Ludvár there were several 2-3 cm thick fish scale layers. According to Trogmayer's (1968) calculations the production of one such layer required the simultaneous processing of something like 800 30 cm long individual fish. He thinks it unlikely that all the meat was consumed immediately; some of it was probably dried and stored for later use.

As to the method by which the Körös people caught most of their fish, net-fishing has been suggested. Large quantities of mostly tomato-shaped objects interpreted as net-weights have been found on the sites (Kutzián, 1944). At Szarvas-Szappanos, for example, 170 such 'net-weights' were counted by Krecsmárik (1915), the excavator of the site. The net-weights were found at different points of the settlement, usually in groups of 40 or 50. Although the interpretation of these clay objects is still open to doubt (Nepper, 1968), it is not unlikely that the Körös people did in fact practice some sort of net-fishing. Net-fishing is usually a communal enterprise in which the male members of several households would join together. (Banner, 1937a).

Another method of obtaining large quantities of fish would have been simply to gather the fish which were left stranded after the floodwaters had receded. According to a sixteenth century description of the Lower Tisza region, there are so many stranded and rotting fish around after inundations, that even the air is polluted by the smell (Oláh quoted by Trogmayer, 1968). In the village of Sára in the Bodrogköz, they used to say half-jokingly that before the regulation of the rivers the villagers had only to open the church doors before the flood came and collect the fish which had filled the church after the waters had gone (Ébner, 1925). There is little doubt that the Körös settlers also took advantage of the abundantly available fish after inundations.

Bone implements resembling double-pointed gorges have also been found on the sites and may have been used for fishing (Trogmayer, 1968). Fish-hooks are virtually unknown, although one has recently been found at the site of Dévaványa-Katonaföldek (Ecsedy, 1972). In view of the zoological and archaeological evidence,it seems likely that the methods most favoured by the Körös people were those which could be relied on to bring in large quantities of fish. It should be pointed out, however, that fishing may not have been of equal importance at every site. At Tiszajenö-Szárazérpart, for example, no fish bones were found although there were some net-weights. On the other hand, only a small portion of the site has as yet been excavated (Selmeci, 1969).

Besides fishing the gathering of freshwater shellfish and snails was also an important element in the Körös subsistence pattern. Banner (1931), the excavator of several Körös and later sites, has observed that while freshwater shellfish and snail shells are generally present at the settlements of the Tisza culture, the shells never occur in such enormous quantities as they do at the Körös sites. At Hódmezővásárhely-Kopáncs tanya, the shell refuse was ½ m thick, and on the neighbouring Tisza site of Kökénydomb the shells were far less numerous. Similarly large shell dumps were also found at other Körös sites: Öbessenyo, Ószentiván, Nagyjaksorpart, Endrod (Banner, 1931), Szarvas-Szappanos (Krecsmárik, 1915), Maroslele-Pana, Gyálarét-Szilági major, Röszke-Ludvar and Deszk-Olajkút (Trogmayer, 1968).

Most of the shells have been identified as belonging to the species Unio pictorum. Less frequent,though usually present,are the shells of U. crassus, U. timidus and Anodonta cygnea. The U. pictorum and A. cygnea live in the calm shallow waters of the floodplain, while the U. crassus prefers the swifter river currents (Czógler, 1933-34). The snail species most commonly found on the sites are Viviparus hungaricus Haz. and Lymnaea stagnalis L. As most of the shells come from the neighborhood of hearths and are usually found mixed with other food remains,it seems likely that they had been collected mostly for their food content. Unfortunately, the amount of food that each shell dump or layer represents has not yet been calculated.

The aquatic environment of the sites was also very rich in birds and large mammals. The analysis of the bone remains from the sites suggests that hunting, similarly to fishing and the gathering of shellfish, was also an important element in the Körös subsistence pattern. At Röszke-Ludvár, 34.07% at Gyálarét-Szilágyi major 27.73%, at Maroslele-Pana 19.26% and at Deszk-Olajkút 6.30% of the animal bones come from hunted mammals and birds (Bökönyi, 1964; 1974). At the Yugoslav Körös sites of Nosa-Biserna Obala and Ludas-Budźak, the figures are 62.7% and 17.67%, respectively (Bökönyi, 1974).

The percentage of bird bones in the total animal bone material varies from settlement to settlement. At Maroslele-Pana it is 10%, while at Deszk it is only 2.6%. At Röszke-Ludvár, Gyálarét and Ludas-Budźak the bird bones constitute approximately 7% of the total. In all cases the overwhelming majority of the bird bones belongs to waterbirds. At Maroslele-Pana,15 species have been identified by Bökönyi. Most of the birds are fisheaters and with one exception they all prefer water-rich habitats. The species are mostly indigenous to the region. Some of the birds, however, may only have spent the winter in Hungary (Bökönyi, 1964). As has been mentioned before, Hungary lies on the annual migration route of many northern and eastern European birds.

Of the larger mammals, the most frequently hunted animals were the aurochs, the roe and red deer, the wild swine, and in some cases the wild ass. With the exception of the aurochs and the wild ass, which prefer the more open habitat of the forest steppe, most of the other animals would have been found in the neighbouring gallery forests. At Nosa and to a lesser extent at Ludas-Budźak, the wild ass bones are fairly numerous, suggesting perhaps a larger degree of dependence on the resources of the forest-steppe than was the case at most other Körös sites. At Röszke-Ludvár the most frequently hunted species was the red deer, at Gyálarét the wild swine. At this stage, however, since the total bone material has been collected and examined at only a very few sites, the extent and nature of variability in the hunting pattern of the Körös culture cannot yet be fully estimated. There is also little indication of the methods of hunting employed by the Körös people. Arrowheads and other projectile weapons are rarely found on the sites. It is possible that most of the Körös hunting equipment was made from less durable materials or that methods not requiring projectile weapons were used (Tringham, 1971). The domesticated dog is present at most sites,and one of its uses may have been for hunting.

On the basis of avifauna remains and the age distribution of the mammalian bones, Bökönyi (1972) has been able to demonstrate that occupation at the sites of Ludas-Budžak, Maroslele-Pana, Röszke-Ludvár and Deszk-Olajkút was uninterrupted all through the year. Although there is no further detailed evidence, by analogy it may be supposed that most of the other Körös sites were also occupied all the year round. As to the seasonal pattern of exploitation, it has been suggested that fishing was mostly practiced in summer, the hunting of forest animals in autumn and winter, and the gathering of shellfish in spring. This suggestion is based on the evidence of a pit excavated at Ludvár which contained stratified deposits of shells, bones of forest animals and fishbones (Trogmayer, 1968; Tringham, 1971).

Evidence for the exploitation of wild plant resources is as yet limited. Acorns and beechnuts were found in storage pits at Nosa and Ludas and had probably been used to supplement the diet (Garašanin, 1961). At the more recently excavated site of Méhtelek, charred hazelnuts were found in a pit (Kalicz and Makkay, 1974). Apart from providing food, wild plants also had other uses. Vegetable matter was usually mixed with the clay by the Körös people in making pottery. They also made rush mats, imprints of which have survived. The trees of the gallery forest were used for house construction and probably also for fuel.

For a long time the Körös culture was regarded as basically hunting-gathering and stockbreeding culture (Banner, 1931; 1937; Kutzián, 1944). In recent years, however, remains of cultivated plants have been recognised at several sites. At others, the existence of cultivated plants has been confirmed through the identification of their imprints in pottery. Wheat remains have been found at Hódmezövásárhely-Gorzsa-Kovács tanya, barley remains at Tiszajenö-Szárazérpart (Selmeczi, 1969), and wheat and barley remains at Szarvas-Szappanos (Hartyányi, Nováki and Patay, 1968). Millet has been found in storage pits at Nosa (Garašanin, 1961). Imprints of einkorn and emmer wheat have been identified in pottery sherds at Röszke-Ludvár and Gyálarét-Szilágyi major. At the latter site, imprints of Bromus and Lathyrus sp., which usually occur as weeds of cultivation, have also been identified (Hartyányi and Nováki, 1971). It is possible that flax may also have been cultivated (Trogmayer, 1968). In contrast with the earlir view, today there seems to be little doubt that the Körös settlers were also agriculturalists.

From the location of the settlements it is possible to draw certain tentative conclusions about the land use pattern and agricultural practices of the Körös people. As we have seen, most of the sites were located on levees immediately by the rivers. The basins which surrounded the levees were inundated through most of the year. With their heavy, badly drained and often alkaline soils, they would have been unsuitable for cultivation. The most likely area for the location of the fields was on the levee tops and slopes. Ethnographical evidence confirms this hypothesis. In those areas of the Alföld where flooding was a problem before the regulation of the rivers, the fields were usually located above the floodplain on elevated ground (Maksay, 1971; Halmos, 1956; 1957). Levees as areas of crop production had a number of advantages over the basins. They were usually safe from flooding; had light easily worked soils and favourable drainage conditions. High ground water table during the critical vegetative period would have ensured good crops even in periods of drought. Besides being easily worked, levee soils were also very fertile. It is possible that the location of the Körös sites in a long narrow stretch along the rivers was connected not only with the presence of the rivers, but also with that narrow strip of fertile soils which today are recognisable as meadow chernozems or chernozems, and which follow the rivers as do the settlements in a narrow ribbon-like formation. The width of these chernozem bands is usually no more than 100 to 200 m across, depending mainly on the size of the levee. The location of the settlements on levees in the floodplain meant that good agricultural land in the immediate neighbourhood of the sites was relatively limited in extent.

Ethnographic data suggests that under similar circumstances, that is when good dry land is limited, fields tend to be relatively permanent. In the Bodrogköz, before the regulations, not only were the cultivated plots always located in the same place, on relatively small expanses of higher ground, but even the crops grown in them were always the same. The name of the crop became associated with the land it was always grown on: field names such as Wheat Field, Hemp Field, Millet Field, etc., are common in the Bodrogköz and also in the Rétköz (Ébner, 1925; Kiss, 1929). There is a Körös site located in a place called Buzás-sziget (Wheat Island) near Tiszainoka, in the Tiszazug, and similar names, preserving the memory of old practices, are no doubt numerous in most regions of the Alföld.

While there is no direct evidence for the permanence of the Körös plots, the depth of the cultural layer at most Körös sites suggests some permanence of occupation. The depth of the Körös settlements is in marked contrast to the shallowness of most ALP sites. The thickness of the occupation deposit at Hódmezövásárhely-Vata tanya is 0.75 m (Banner, 1935), at Szelevény-Kisasszonypart I about 1.5 m (Kalicz, 1957), and at Hódmezövásárhely-Zsoldos tanya between 1.32 and 3 meters (Banner, 1932). It is perhaps of some interest that at the medieval site of Móric also in the Alföld, where the same building material was used, the thickness of the occupation layer is approximately 0.5-1 m, and the lifespan of the village was according to documentary evidence over 75 years (Méri, 1954). By analogy it may be supposed that the Körös sites were occupied for a similarly long period of time.

Permanence of settlement implies at least the possibility of the existence of permanent 'fields'. In the case of the Körös culture, heavy reliance on the rich aquatic resources of the area may have contributed to the stability of the settlements. Nevertheless, agricultural production must also have been sufficiently profitable to support occupation in the same place for a relatively long period of time. Local soil exhaustion need not have been an insurmountable problem. Levee soils, occasionally coated by inundation silts were naturally fertile. There were also other methods of replenishing soil nutrients which were within the technical capability of the Körös people. Clark (1947) notes the increase of sheep in importance in Bronze and Iron Age Europe as settled agriculture with fixed fields became the usual practice. The fallow provided an ideal feeding ground for the sheep, while the sheep provided manure for the fields. In Hungary, even today, sheep are folded not only on the fallow, but also on the cultivated fields (Gombás, 1960). From early spring until the end of April the sheep are turned out onto the fields to thin out the sprouting crops to prevent the plants running wildly to leaf. In autumn, following the harvest the sheep are grazed on the stubble. Even in winter, they are occasionally let out onto the fields to break up the ice. Similar practices have been described from Greece, where no other manure than that from sheep grazing over the sprouting crops and stubble is used (Renfrew, J.M., 1972). The antiquity of such practices is probably great and it may even reach back into prehistoric times. It may be remembered that in the Körös culture sheep were the most important domesticated animal.

The settlements of the Körös culture were not, however, occupied more than once; there are no Körös tell sites. One reason why the sites were not reoccupied may have been the increase in the danger of flooding. Even small changes in the drainage pattern of the area could have been disastrous in the life of the settlement. Nevertheless, it is more likely that cultural factors were as or more important in the mobility of the sites than purely environmental ones. As yet there are no reliable figures on Körös or later population densities from the area: the lack of tell sites suggests that during the life of the Körös culture there was no shortage of suitable land where new settlements could be founded if the old sites became undesirable.

The seasonality and vehemence of the floods must, on the other hand, have been important in influencing decisions on crop policy. Due to the location of the sites, destruction of the crops by floods was an ever present danger. Because of the risk involved, it is unlikely that the Körös farmers would have relied on a single crop or even a single area of cropping. A flexible crop policy relying on a variety of cultivated plants would at least ensure that some crops were harvested each year. Einkorn wheat, which is poorly suited to irrigation agriculture (Butzer, 1965), was probably mostly grown on the levee tops. Emmer and barley may have been grown on both low and high lying ground. It is not yet known whether the cereals cultivated by the Körös people were autumn or spring sown. If the *Bromus* species found at Gyálarét was *Bromus secalinus*, it would suggest autumn-sowing. *Bromus secalinus* is a common weed of cultivation, occurring mostly with winter wheat (Tempir, 1964). Beside wheat and barley, millet was also grown by the Körös people. Millet has a very short growing season and is therefore well adapted to flood plain agriculture (Kiss, 1929). In the Alföld, it is usually sown in late spring or early summer after the floodwaters have receded. Thus, if all other crops failed, millet could still be relied on to bring in a harvest. Körös agriculture may have been precarious, but it could produce at least one crop a year.

Beside farming, stockbreeding was also practised in the Körös culture. While good agricultural land was relatively limited in the neighbourhood of the sites, land suitable for stock rearing was more plentiful. In the animal bone material found on the sites, sheep is represented by the highest percentage among the domesticated animals. Calculated as percentages of the total number of animal bones, the figures for sheep/goats vary between 30 and 68% (Bökönyi, 1974). Cattle seems to have been the second most important domesticated animal in all the Körös sites. Cattle bones constitute 7 to 22% of the total bone material. The domesticated pig, though present at most sites, was of relatively little importance. The proportion of pig bones is usually less than 3%. The domesticated dog is represented in similarly small numbers.

Bökönyi attributes the predominance of caprovines to the influence of Near Eastern cultural tradition. It is possible, however, that their importance was at least in part due to local conditions of alkali soils and vegetation in the vicinity of the sites. The majority of the Körös domesticated animals seems to have come from foreign stock, although there is also some evidence for the local domestication of the indigenous aurochs population (Bökönyi, 1964).

In temperate Europe, in contrast to the Mediterranean, an important limiting factor to stock breeding was the scarcity of the winter fodder. Conditions in the Alföld resembled in this respect the Mediterranean rather than midlatitude Europe. Extensive floodplains and swamps provided until recently adequate winter pasture for the domesticated stock of the region. While Kruk (1971) argues that extensive meadows did not exist in river valleys in temperate Europe in the Atlantic period, the existence of many types of priseral vegetation in the floodplains cannot be disputed. Imprints of reeds and sedges, for example, have been found in Körös and ALP pottery. According to the ethnographical evidence, reeds and sedges were especially important in the wintering of cattle. As these tall plants were not covered by snow, cattle could easily browse on them. For sheep they were less suitable because of the danger of liver fluke (Jankovits, 1967). However, other green pasture was also available in the floodplains and swamps (Glaser, 1939). Floating marshes and small islands were favoured places for wintering stock (Dankó, 1962-63). Before the regulations, large flocks of sheep from Transylvania were often wintered in the Alföld (Gombás, 1960). Besides providing winter fodder, the floodplains were also used as summer pastures. By mid-July, just as the effect of drought was beginning to be felt on the higher loess plateaus, the last of the floods had receded from the floodplains, leaving rich pastures behind (Glaser, 1939).

Grazing schedules in the Alföld followed the same pattern: livestock was kept near the villages till early spring, when pastures on the high loess plateaus became available. From about mid-July till after the harvest, the herds were grazed on the floodplain, and after the harvest on the stubble. In winter, pasture was available only in the floodplains (Glaser 1939). While no evidence of the actual Körös or ALP grazing practices has survived, it may be supposed that the annual cycle followed a similar schedule.

Evidence available at present suggests that the settlements of the Körös culture were more or less self-sufficient. There seems to have been no development towards a functional hierarchy of settlements. Settlement size shows less variation in the Körös culture than in the ALP. Most Körös sites are large in comparison. There is also a certain regularity in the spacing of the sites; distances between settlements tend to be uniform (Kalicz, 1957). Although there was some trade in raw materials, for example in obsidian and other rocks, the volume of trade was relatively small. Obsidian and other stone tools occur only in small quantities on the Körös sites (Kalicz, pers. comm.). At the recently discovered site of Méhtelek, obsidian tools, on the other hand, constitute over 80% of the chipped stone assemblage (Kalicz and Makkay, 1974). Alongside the finished tools, there is also a large quantity of unworked obsidian on the site. Méhtelek is in many ways an exceptional site. Culturally it has more affinities with the Körös group of Transylvania than with the Körös group of Hungary, the main distribution area of which lies more than 150 km southwest of Méhtelek. Furthermore, the material culture of the site has a number of traits which represent a departure from the Körös cultural tradition altogether and connect Méhtelek with the emergent group of the Early ALP.

Alföld Linear Pottery Culture

Much of what has been said about the Körös ecology is also valid for the ALP. The distribution areas of the two cultures overlap, and a large percentage of the ALP sites are found in microenvironments similar to or identical with those of the Körös sites. Over 90% of the ALP sites are located in areas of insufficient humidity, and over 50% of the sites are on soil types where water had been an agent of soil formation. The same environmental limiting factors - insufficient moisture in the growing season, danger of floods and scarcity of good dry land - which applied within the Körös cultural area were also influential within a large segment of the ALP distribution area. There were differences, however. Some of these were due to environmental changes as the distribution area of the Körös culture was extended northwards by the ALP culture, while others were probably due to changes in the pattern of exploitation.

Early ALP

In contrast to the Körös culture, where all the settlements are located in the forest-steppe zone, in the ALP settlements from the earliest phase appear also in the forest zone. The associations most preferred by early ALP settlers were oakwood on sand and loess. There was also some settlement in acidiphilous oak forests. All these associations are characterised by a relatively low density of forest cover: densely wooded areas were avoided. It may be remembered that all the settlements at this stage are still located in areas of limited moisture in the growing season.

The settlements in the forest zone, just like those further south, are on small expanses of higher ground in the floodplain. In the Early Phase, a relatively large percentage (24%) of the sites are found on sand. Although most of the sand in the area is windblown sand, it is worth noting that in the neighbourhood of the sites the sand usually contains some amount of loess (Borsy, 1953).

Although as yet there is no direct evidence to show whether the Early ALP settlers practised agriculture in the forest zone, the nature of the soil types occupied is some indication that agricultural considerations were at least a factor in determining the location of the sites. Settlements were generally located on the most fertile soil types of the area. In the forest zone,86% of the sites are found on brown forest, chernozem-brown forest and kovárvány brown forest soils. As we have seen, the brown forest and chernozem-brown forest soils are the most fertile of all forest soils in Hungary. Kovárvány soils, though less fertile, are exceptionally productive for sandy soils. They also had the added advantage of being easy to work. The occasional rise of birch pollen in the diagrams of the Nyirség may be interpreted as further evidence for the presence of agricultural man in the area. Birch is usually the first tree to appear after the forest had been burnt, and regeneration is allowed to take place (Iversen, 1956). The pollen diagrams also show the presence of cultivated plants in the area in the Atlantic period, but in the absence of radiocarbon dates it is difficult to know with which phase of the ALP their first appearance may be correlated. It is not known either how permanent agriculture was at this stage: the small size of the settlements and the shallowness of their culture layer would suggest that occupation at the sites was relatively short lived. It is perhaps of some relevance that the vehemence of the floods along the Upper Tisza and Szamos rivers, where most of the early sites are located, is generally much greater than along the lower reaches of the Tisza (Andó and Vágás, 1972). Before the regulation of the rivers, most cereals were spring sown in the area, partly to avoid flood damage, and partly because winter wheat would have been decimated by rust (Kiss, 1929).

Evidence about Early ALP stockbreeding is also very scant. At the Szatmár site of Tiszacsege,sheep is still the most important domesticated animal (65%). Cattle bones constitute 15% of the domesticated animal remains;and pigs 10% (Kalicz, pers. comm.). The composition of the herds in the forest zone was probably somewhat different although already at Tiszacsege the slight rise in the importance of the domesticated swine would indicate a growing dependence on woodland resources.

One feature that is shared by all the early sites is the great abundance of obsidian in the lithic material. At Tiszacsege, for example, all the tools are made of obsidian (Kalicz, pers. comm.); at Méhtelek 80%. As obsidian was very highly valued in the neolithic, the deposits in the Tokaj Mountains, one of the richest sources of obsidian in southeast Europe, must have exerted a great influence on the location of the sites. The attraction of the obsidian source and the presence of other raw materials in the mountains may in fact have been the reason why the distribution area of the ALP was extended into the forest zone in the first place. It is possible that some of the early settlements functioned primarily as 'trade-posts' whose main task was to obtain and transmit valuable raw materials to sites further south.

Middle ALP

There are no marked changes in the settlement pattern: settlements occur both in the forest-steppe and the forest zone. As in the earlier phases, the majority of the sites are located in azonal situations, in or near the edge of the floodplain. At most of the sites, humidity is still insufficient in the growing season, although the moisture deficiency is less severe than in the Körös culture.

The culture layer at the settlements is usually thin and consists of no more than one habitation level, suggesting that occupation at any one site was of relatively short duration. It is possible that some of the sites, especially in the mountains, were occupied only on a seasonal basis. At Uppony-Malomgát, Korek (1961) excavated two circular hut structures. The circular house form is unusual in the ALP, where, just as in the Körös culture, the typical house is rectangular. The site at Uppony may have served as a temporary shelter for seasonal herdsmen (cf. Flannery, 1972). The colder winter and shorter growing season may have discouraged year round settlement in the mountains although the Northern Mountains in Hungary are not high enough to be really inhospitable. It should also be pointed out that there is usually some good,if limited,agricultural land in the river valleys in the vicinity of most sites,so the possibility that agriculture was also practiced cannot be excluded.

Settlements in the Middle Phase are again mostly small. The smaller size and greater mobility of the early and middle ALP sites are in contrast with the larger size and greater permanence of the Körös sites. As many of the ALP sites are found in the same microenvironments as the Körös sites, environmental change alone cannot be held responsible for the difference. It is more likely that the difference in settlement size and permanence was due to changes in the pattern of exploitation of the environment. Increased reliance on cattle and swine may have been part of the change. Unfortunately, very few ALP sites have been excavated so far, and the faunal and botanical evidence is still very fragmentary. The impression is that the Middle Phase was still largely a phase of experimentation with local resources and with different ways of exploiting these resources.

Late ALP

The Late Phase is characterised by the development of a number of regional groups. These groups have been distinguished on the basis of pottery styles, but the environmental analysis of the sites suggests that their ecologies were also different. Over 50% of the sites in the Tiszadob group, and over 60% in the Bükk group are located in the forest zone; the majority of the sites in both groups are found on forest soils. In the Esztár group, the largest percentage of the sites are found in the moderately dry forest-steppe; the majority of the sites on soil types where water had affected soil formation. The least homogenous of the groups are the Late AVK and Szilmeg: settlements occur in more or less equal proportions in the dry and moderately dry forest-steppes and in the forest zone. The Szakálhát group, whose sites are almost exclusively in the dry forest-steppe, seems to have more in common with the Körös culture as far as relationship to the environment is concerned than with the rest of the ALP groups.

With the emergence of regional groups came a new development toward greater functional diversity in eastern Hungary. Specialized sites and tell settlements appear in Hungary for the first time. Also, for the first time there are indications of the development of an incipient settlement hierarchy in this area. Accompanying these advances, were changes in the pattern of farming and stockbreeding.

Specialized settlements are mostly associated with the Szilmeg, Tiszadob and Bükk groups. At Boldogköváralja-Tekeres Patak (Bükk I) 6 small rectangular houses and adjoining the houses 4 work floors were uncovered (Kemenczei, 1964). The work floors were littered with raw materials, half finished and finished tools. From beside one of the houses came a small pot containing 567 flint blades. At the Szilmeg site of Eger-Kiseged 300 polished stone axes were found (Galván, 1958). At the sites of Onga (Tiszadob, Bükk), Megyaszó (Tiszadob) and Edelény-Borsod (Bükk III), there were also large quantities of stone tools, suggesting intensive manufacture (Kalicz and Makkay, 1977).

It is unlikely that such large quantities of tools would have been produced for home use only. According to Vértes (1965) the flint blades found at Boldogköváralja are remarkably homogenous, both typologically and functionally. He suggests that they were probably intended for sale or barter. Increased production is not the only evidence that the volume of trade was increasing. Obsidian, which must have come from the Tokaj Mountains, was becoming more abundant at most ALP sites and also outside the ALP distribution area. At Vinča, obsidian is insignificant in the early A phase, but is fairly common in the Later B phase (Csalog, 1955), which parallels the Late ALP. Tiszadob, Bükk and Szakálhát sherds have been found as far north as Poland and as far south as Thessaly (Kalicz and Makkay, 1977). Szakálhát sherds are especially numerous in the Morava valley of Yugoslavia. One item of exchange which was carried north from the Mediterranean was Spondylus. Spondylus beads have been found at Istálloskö and Nagykálló among other sites. What other substances were traded beside obsidian, flint and Spondylus is not known. Nor do we know how the trade was organised or what the mode or modes of exchange were. It would be interesting to know, and hopefully further research will be able to show, whether the Spondylus found on the TLP sites came directly from the south or via the ALP culture; or what substance or commodity was traded in those Notenkopf and Zseliz pots which were found on the ALP sites.

Beside external trade, internal trade was also increasing. Pot sherds of one group are often found on sites of another group. It is possible that some of the pots contained foodstuffs which would imply that food production had reached a higher than mere self-sufficiency level. It may also be assumed that the exchange of food products would be paralleled by a greater regional diversity in food production. Although there is some faunal evidence, discussed later, to support this view, the data is too fragmentary to allow a positive conclusion.

Although not necessarily conclusive by itself, the appearance of tell-like settlements in the Szakálhát group is also suggestive of more intensive agricultural production. At Battonya-Gödrösök, the site has 2 building horizons. At Tápé-Lebö-Felsöhalom, there are probably 3 Szakálhát levels; their thickness is 2 m. The site of Dévaványa-Sártó was also probably rebuilt more than once during Szakálhát times; the thickness of the cultural deposit is 140 cm.

The argument for more advanced agricultural production is strengthened by the available botanical evidence. It seems that at least two major changes occurred in the farming pattern at this time. One was the increase in the importance of barley. A sizeable quantity (c. 1000 cl) of six-row hulled barley has been found, for example, at the Szakálhát site of Dévaványa-Réheligát (Hartyányi and Nováki, in press). The find comes from two collapsed houses, from among the remains of large pottery storage jars. The Réheli gát find is the first evidence of six-row barley in Hungary. The second change was in crop purity. At Réheli Gát, in House II, one storage jar contained 20 cl six-row barley, another one 250 cl barley and a third one, also in the same house, approximately 1500 cl emmer wheat. Mixed with the emmer, there were also three grains of einkorn wheat, 5 grains of lentil and a very small number of weed grains. The barley finds were completely pure. The separate storage of wheat and barley suggests that they had been sown and harvested separately (Hartyányi and Nováki, in press). The small quantity of weed grains found with the cereals argues for good weed control. 'Increasing crop purity represents increasing specialization and a vastly improved agricultural competence. It opens up possibilities of further sophisticated advances such as the rotation of crops (eg. cereals then pulses) to regenerate the land' (Renfrew, 1972, 276). Whether crop rotation was an established practice in Hungary at this time is not known. As at least one of the pulses was known to the Szakálhát people, the possibility that some form of crop rotation was practised cannot be excluded.

What is interesting about the advances that were made by the Szakálhát people is that they were not an isolated phenomenon. They, in fact, closely parallel developments in Late Neolithic Greece (Renfrew, J.M., 1973). Although the botanical evidence is limited, similar changes toward greater agricultural competence may be inferred on archaeological grounds in most other areas of southeast Europe.

The evidence of soil types also supports the argument for increased agricultural production in the Szakálhát group. Although the Szakálhát settlements are found on the same soil types as the Körös sites, a higher percentage of the Szakálhát sites are located next to larger tracts of fertile meadow chernozem soils.

It is not impossible that the plough was already known at this stage. The same suggestion has been made for Late Neolithic/Chalcolithic Greece (Renfrew, 1972). The increase in the importance of cattle in Hungary at this time and the agricultural developments outlined above would be consistent with the use of the plough. Although there is no direct evidence that the plough was known in neolithic Europe, there is evidence for the presence of oxen in the Linear Pottery culture (Murray, 1970). The oxen would have provided the necessary draught power for pulling the plough. Zoological evidence at two sites in the Vădastra culture of southwest Rumania and northwest Bulgaria indicates that cattle were in fact used for traction (Ghetie and Mateescu, 1973). The sites are located near large expanses of slightly leached chernozem soils. The large quantities of cereal remains which were found on the sites would suggest intensive agricultural production: one pit alone contained 1600 kg wheat. The Vădastra culture was partly contemporaneous with the Late Phase of the ALP.

While there are no tell settlements outside the Szakálhát group in Hungary, stratified sites with thick occupation deposits occur in other Late groups. At Boldogkőváralja-Tekeres patak, for example, the settlement consists of 2 habitation levels and the occupation layer is 160 cm thick. At the Tiszadob site of Sajőkeresztur-Homokbánya, the cultural deposit is 70-80 cm deep, and at Borsodbóta the depth of the Bükk occupation layer is 150 cm.

In stockbreeding,the development was toward increased reliance on locally available species. Cattle bones constitute 40 to 50% of the bone remains. Alongside with cattle, pigs also rose in importance. In contrast to the Körös sites, where pigs were almost negligible, on the Late ALP sites they represent 10 to 30% of the total bone finds. Cattle are well suited to the forest-steppe and pigs to woodland environments. Wild species of both animals are present in Hungary.

Reliance on three (cattle, sheep, pigs) instead of two species (sheep, cattle) as was the case in the Körös culture meant more flexibility and allowed for greater diversity between sites in the composition of the domesticated stock. While at Tiszavasvári-Keresztfal sheep was the second most important domesticated animal, at Polgár-Folyás-Szilmeg pigs were more important than sheep. At Edelény-Borsod, Tápé-Lebő-Felsőhalom and Dévaványa-Sártó sheep and pigs were of more or less equal importance. The differences to some extent reflect the immediate environments of the sites: Tiszavasvári is located near the loess steppe of Hajduság, while the other sites are surrounded by extensive gallery forests.

Although sheep and pigs were both important in the stockbreeding of the ALP, cattle was invariably the most important domesticated animal on the sites. Unlike sheep or pigs, cattle require considerable pasturage. According to ethnographical data from the Alföld the pasture requirement of 1 bovid equals that of 5 sheep or 5 pigs (Györffy and Viski, no date). If we assume that the total number of domesticated animals were the same on the Körös and ALP sites and that different areas were used to pasture the different species, the size of pasture needed to feed the domesticated animals of the ALP would have been almost three times as great as that needed for the animals of the Körös sites. (The calculation is based on the percentages of individual animals found at the Körös sites of Gyálarét, Röszke and Ludas, and at the ALP sites of Edelény-Borsod, Folyás-Szilmeg and Tiszavasvári-Keresztfal). The ALP solution seems to have been to reduce the size of the settlement unit and thus presumably the number of animals which needed to be pastured. A large percentage of the ALP sites are so small that they could not have accommodated more than a very few families (Kalicz, 1957). It may also be argued that the increase in pasturage requirements consequent on the change in the composition of the stock was at least in part responsible for the greater mobility of the ALP sites. With the exception of the tells and stratified settlements of the Late phase, the ALP sites are generally shallower than the Körös sites, which suggests that occupation on them was short lived.

Hunted animals constitute on the average 20% of the faunal remains. At Lebő the percentage is much higher, but there no separation has been made between the Körös and Szakálhát material. It is interesting to note that the number of fishbones is usually very small on the ALP sites. As cattle supplies more meat than sheep per animal, it is possible that with increased reliance on cattle,fish became less important as a source of protein. However, it should be remembered that the bone material, at present, comes from only a very few sites.

Kalicz (1957), in his survey of the Tiszazug region, calls attention to a number of interesting contrasts between the settlement patterns of the Körös and ALP cultures: 1) The number of ALP sites mostly(Szakálhát) is much larger,and the size of the settlements is much smaller than in the Körös culture. 2) Not all the ALP sites are of equal size. While the majority of the settlements is small, there are also 4 sites whose dimensions are comparable to those of the Körös sites. 3) The sites are not evenly spaced; there are generally 3 or 4 smaller sites next to one another in a group. Similar patterns of distribution have also been observed near Hódmezővásárhely.

Locational studies of modern settlements suggest a positive correlation between settlement size and functional range (Haggett, 1965). By analogy, it may be supposed that the same relationship also obtained for the Szakálhát settlements. Unfortunately, none of the sites had been excavated yet, so the hypothesis is unsupported by archaeological evidence. But together with other developments,such as the appearance of tell sites, increased trade and more intensive agriculture, the existence of a simple form of settlement hierarchy does not seem unrealistic. It should not be forgotten that the Szakálhát group is the southernmost group of the ALP with strong ties with the Vinča culture, where similar developments were presumably taking place at the same time.

The multiplication of settlement units, the appearance of tells and increased specialization would suggest that population increased during the Late phase. However, population estimates are notoriously difficult, and as we have seen, the majority of the sites is smaller than the average Körös site. Nevertheless, it is interesting to note that toward the end of the Late phase settlements for the first time appear on hilltops in defensible position. Similar developments have been observed in other parts of southeast Europe, where increased population density and consequently greater competition for good agricultural land have been cited as the main reason for the shift in settlement pattern (Trbuhović, 1968; Sherratt, 1972). It may be remembered that the northern mountain region in Hungary was also a source of valuable raw materials ; competition here may have been just as much for access to raw materials as for good arable land.

Transdanubian Linear Pottery Culture

Although in the ALP almost 30% of the sites are located in the forest zone, over 90% are still in areas where climatic moisture is limited in the growing season. In the TLP, on the other hand, 50% of the sites are in areas where rainfall exceeds evapotranspiration. In view of this, it is hardly surprising that the majority of the TLP settlements (67%) are found in the forest zone. With increased settlement in more humid regions, the importance of the floodplain in the ecology of the TLP sites decreases. Within a 2.5 km radius of the sites only 25% of the land is floodplain or marsh. In other words, in the TLP there is a shift toward increased exploitation of the zonal woodlands. The economies of the Körös and the ALP cultures were, as we have seen, largely dependent on the azonal environment of the floodplains. Naturally, there is a fair degree of variation in the ecological pattern of the TLP sites; some settlements relied more on the resources of the floodplains and marshes than did others.

Early TLP

The locational pattern of the Early Phase differs in some respects from that of the fully developed TLP. It may be remembered that the pottery styles of the Medina Horizon have closer links with the Starčevo culture than with the Linear Pottery culture. In the ecology of the sites,there are certain similarities to the Körös culture although differences exist even at this early stage in the development of the TLP.

The Körös sites are, as we have seen, located in the Alföld, where moisture is deficient through most of the year. While the majority (86%) of the early TLP sites are also in areas of insufficient moisture, some settlement occurs in more humid regions. Moreover, the degree of aridity is much less severe in Transdanubia than it is in the Alföld.

While the early TLP sites are, similarly to the Körös sites, located on levees or loess covered dunes in the floodplains or on lake shores, aquatic resources seem to have been of much less importance than in the Körös culture. At Aba-Ángyihegy, for example, fish and water birds are very rare in the bone material, although the site is situated on a levee in the middle of a 6-7 km wide floodplain (Makkay, 1970). On typological grounds, the site of Aba-Ángyihegy is somewhat later in date than the other early sites.

The early TLP sites are usually smaller and have shallower cultural deposits than the Körös sites. It is interesting to note that so far only circular hut structures have been found on the sites of the early phase. Small circular huts have been uncovered at Bicske (Makkay and Kalicz, 1972), Medina and the somewhat later site of Kiskánya (Csalog, 1941). It is possible that some of the sites were occupied only seasonally. But whether occupation was on a seasonal basis or all the year round, it lasted only for a short period of time at any one site. Unfortunately, in the absence of botanical and faunal data, no real attempt can be made at the reconstruction of the early agricultural and stockbreeding patterns. At present, the number of early sites is so small that the locational pattern itself is rather tentative.

Middle TLP

The Middle Phase may be distinguished as the first phase of the fully developed Linear Pottery culture. Settlements of this phase are found throughout the loess lands of central and also parts of eastern and western Europe. So far, no settlements belonging to this phase have been found in France, Holland, Belgium and the Ukraine (Pavúk, 1973). Pottery styles show a marked uniformity from one end of the distribution area to the other. Ecological conditions are also apparently similar at most Linear Pottery sites (Tringham, 1971). The settlements are located on loess in light mixed oak forests.

In Transdanubia, 77% of the sites are in the forest zone; the rest of the sites are in the moderately dry forest-steppe. The preferred associations in western Hungary are also oak forests: acidiphilous and hornbeam-oak. (The hornbeam was probably rarer in the oak forests of the Atlantic period than it is today.) The Hungarian pattern differs slightly from the general one in that fewer sites (42%) are located on loess. The majority of the settlements are, however, on well drained parent materials.

Botanical and faunal evidence is again limited. Nonetheless, from the preferential location of the sites on the more fertile soil types, it may be inferred that agriculture was an important component of the economy of the TLP. As we have seen, the most frequently occupied soil type is the brown forest soil, the most productive soil type among the forest soils. The settlements are generally located at the junction of two or more ecological zones; most commonly, they are situated on wooded hillsides above the wet floodplain or marshy lakeshore. In Germany, Sielmann (1972) observed the same relationship between site location and topography; the majority of early Bandkeramik sites are found on high ground, usually on the upper part of the slope which rises above the river.

Ethnographical data suggest that the most likely location for the cereal fields of the TLP would have been on the slopes. Horváth (1908), describing the settlement geography of the Zalavölgy, observes that the village wheat fields are generally located on the slopes and the vegetable gardens on the alluvium. The same pattern of land use has been noted by Wallner (1926) in the region of Felsőőr, by Király (1963) in the Kapos valley and by Gállos (1960) in the Mecsek region. The observations refer to both relatively recent and older medieval practices.

Hillsides in Transdanubia, where moisture is less limiting than in the Alföld, offer a number of advantages for agriculture. The slopes, especially when facing south, receive more insolation than the lower lying river valleys or lake shores. As mentioned before, where the topography is varied, braunerden are usually found on the slopes, while the higher plains are covered by less fertile lessivated brown forest soils. The soils in the river valleys are generally too heavy for easy cultivation. In the loess regions, the so-called 'slope-loess' often contains buried soil horizons. Soils developed on such deposits are more productive due perhaps to better physical properties. In addition, drainage conditions are usually more favourable on the slopes. As a rule, the upper slopes are drier, and the foothills where the run-off collects moister. By cultivating the whole face of the slope, TLP farmers would have been able to reduce the risk of crop failure caused by minor fluctuations in precipitation. While in dry years, crops may have failed on the upper slopes, the moister foothills would have ensured a reasonably good harvest. Conversely, in wet years, crops may have been destroyed by rust and rot at the bottoms, but better yields on the upper slopes would have compensated for the loss (Boros, 1971). If land on the slope was insufficient to supply the needs of the village, the arable could be extended to include the less fertile soils of the interfluve. It is possible that corresponding to the difference in soil fertility there were two types of land use. The slopes and, where conditions permitted, parts of the alluvium may have been cultivated continuously and the higher plains only temporarily. The existence of continuously cultivated plots does not necessarily imply permanent settlement. In tropical regions, where ethnographical data are more readily available, a shifting settlement pattern is often associated with continuously cultivated fields.near the settlement and less intensive land use further away from the habitation. The question of shifting cultivation in the Linear Pottery culture will be discussed later in greater detail.

Forest clearance at the technological level of the Linear Pottery culture must have been by slash-and-burn (Clark, 1952). After clearance, the first crop sown was probably millet. In the ethnohistorical literature of central Europe,millet, which has a short growing season and gives exceptionally high yields on newly cleared land, is usually referred to as a 'clearance-crop' (Takács, 1967). The imprint of a grain of Panicum miliaceum L. has been identified by Füzes on a pottery sherd found at the site of Adorjánháza (Bakay, et al., 1970). Wheat seems to have been the main crop, however. At the site of Pári-Altacker, one Linear Pottery pit contained about 5 litres of wheat (Torma, 1970). The find consisted mostly of emmer wheat, although einkorn was also represented by a few grains (Hartyányi and Nováki, pers. comm.). Bread wheat and barley were found at Zánka (Hartyányi, Nováki and Patay, 1968). The presence of Agrostemma githago L. at Zánka indicates that some of the wheat grown by the TLP people was autumn sown (Tempir, 1964).

The only site where faunal remains have been identified is Zalavár (Bökönyi, 1974). Here the number of wild animals outweighs that of the domesticated ones. Aurochs and roe deer seem to have been the most important among the hunted animals; the only domesticated species were cattle and pig. Of the two, cattle is the more frequent. It is probable that cattle was the most important domesticated animal at most TLP sites, with pigs coming second at least in the more humid forest region. On the other hand, it is more likely that on the dolomite hills of the Bakony Mountains or at the margins of the forest-steppe sheep would have been more numerous than pigs.

What the typical house type was in this phase is not yet certain. The circular hut form has survived from the early phase, as their presence at the sites of Balatonendréd and Bonyhád would indicate. The house form at Szentlőrinc-Téglagyár was rectangular with rounded corners. The cultural layer at most sites is shallow, suggesting that occupation was relatively short lived. However, there are exceptions. At Sármellék, for example, Pekáry observed several habitation levels (Bakay, et al., 1966), and at Balatonszabadi-Sáfránykert the thickness of the Linear Pottery occupation was 80-100 cm (Kocztur, 1964). The size of the settlements varies. It is possible that some of the smaller sites, especially those in the mountains,were occupied on a seasonal basis. The settlements were probably self-sufficient, except perhaps for some local trade in raw materials. There is evidence for some mining and manufacturing activity by the inhabitants of the sites of Mencshely-Murvagödrök and Vörösstó-Ragonya. Contact with neighbouring cultures seems to have been limited although sea-shells and some Vinča pottery occur occasionally on the sites.

Late TLP

In the Late Phase there is a definite shift in the settlement pattern. While in the Middle Phase over 60% of the sites were located in moderately humid regions, in the Late Phase the majority (75%) of the sites is found in areas of limited climatic moisture. As may be expected, with the increased occupation in the drier regions there is an increase in the number of settlements located on alluvial deposits. As to soil types, the change is not very pronounced: most of the sites are still found on forest soils, although the percentage of sites on chernozem-brown soils is higher in this phase. The driest chernozem areas are on the whole avoided.

According to faunal evidence, reliance on domesticated resources increased in the Late Phase. Of the early and middle neolithic cultures of Hungary, the Late TLP has the smallest percentage of wild animals: 8 to 13% of the total faunal remains. Fishing, as attested by the presence of large catfish bones, was part of the subsistence economy. Some of the fish must have weighed over 100 kg (Makkay, 1970). However, in the TLP aquatic resources never attained the same importance as they did in the Körös culture. At Neszmély-Tekeres patak,fish bones constitute only 4% of the total bone remains, while at Győr and Pomáz no fish remains are mentioned in the faunal analyses (Bökönyi, 1959; 1974).

Among the domesticated animals cattle was the most important species. Sheep were second in importance,and pigs third. With the exception of the East German sites, ovicaprids are almost absent in the Linear Pottery settlements (Tringham, 1971). The relatively high percentage of sheep/goats in north-west Hungary may have been a response to the drier forest steppe conditions which existed in the neighbourhood of the sites.

Botanical evidence about the agriculture of the Late Phase is limited. So far, only a few barley grains from the site of Győr have been identified. Nonetheless, archaeological evidence would seem to indicate increased agricultural production. At Sukoró-Tóradülő, at least 8 querns were found in an open space between two long-houses. At the site, which consists of two rows of long-houses, there is some evidence for stratification. In Makkay's (1970) opinion the site was occupied for about 80 to 100 years. There are also at least two building phases at Győr-Pápai vám, and pottery styles indicate a relatively long period of occupation at Bicske and Mende. On the whole, the settlements of the Late Phase appear to be more substantial,with their long houses and thicker occupation horizons. There are exceptions, however; at Pomáz, for example, a circular hut has been excavated in a relatively thin culture layer. The circular hut form is also known from Békásmegyer.

It may be remembered that in the Late ALP settlements appear on hilltops toward the end of the phase. In the Late Phase, TLP sites also begin to occur at higher elevations. The highest TLP site, Szentjakabfa-Lázföld, is situated 340 m above the present day village. At Békásmegyer-Vöröshadsereg utja,the settlement is surrounded by a ditch (Csánk, 1964). The function of the ditch is not known, but its presence would suggest some territorial awareness on the part of the occupants of the site.

Notenkopf and Zseliz style sherds are often found on the settlements of the Late ALP, and Szakálhát and Bükk sherds on the settlements of the TLP. Contact with the Vinča culture is also indicated by the pottery finds. In general, influences from the south are more apparent in Transdanubia than in the rest of the Linear Pottery distribution area. One instance of southern influence is the presence of painted pottery in the Zseliz group. Painted pottery is generally absent in the Linear Pottery culture. Another example is the occurrence of clay spindle whorls and weights in Transdanubia. These objects, found, for example, at Sukoró, are mostly lacking in the more northern Linear Pottery assemblages. Their presence in Hungary may be associated with the greater importance of sheep and goat breeding as suggested by the bone remains.

The Linear Pottery culture is considered to be the best known example of shifting agriculture in neolithic times. In Clark's (1952, 96) words, the very rapidity by which the settlements of this culture appeared over the loess regions of midlatitude Europe 'is consistent with the notion of peasants clearing small patches of forest, taking a few easy crops, and passing on to fresh grounds'. Settlements such as Köln-Lindenthal in Germany, Brzesc Kujavski in Poland, or Bylany in Czechoslovakia were abandoned and reoccupied several times. According to Soudský (Soudský and Pavlů, 1972), there was a deliberate pattern of cyclic movement at Bylany, in the vicinity of which 3 to 4 Linear Pottery settlements have come to light. In Soudský's opinion, the initial site would be occupied for 10-15 years, after which it would be abandoned to allow the soil and the forest to regenerate. The settlement would be moved to a new site, which would in turn be abandoned after 10-15 years. Every 30-50 years the old site would be reoccupied and then again abandoned after 15 years. At the settlements of Elsloo, Sittard and Stein in the Netherlands, on the other hand, there seems to have been no breaks in the occupation. According to Modderman (1970), these sites were occupied continuously for over 300 years. According to Kruk (1973) and Wiślański (1970), the immediate vicinity of the sites were more intensively cultivated while the slash-and-burn technique (implying shifting cultivation) was of relatively little importance.

Direct evidence for the agricultural system of the Linear Pottery culture is as yet lacking. Archaeological evidence would seem to indicate a good deal of variation in the permanence of the settlements. The relative thinness of the culture layer of most Linear Pottery sites together with the frequent abandonment of the settlement would indicate that shifting cultivation involving a shifting settlement pattern was the general rule. On the other hand, the available evidence also suggests the existence of permanent settlements and by implication the practice of more permanent agriculture. Although as Carneiro (1961, 50) has pointed out 'the mere fact of practicing shifting cultivation does not necessarily prevent a society from maintaining an essentially sedentary community', permanent settlements are often an indication of more permanent agriculture. Permanent agriculture in the case of the Linear Pottery culture need not imply the existence of an elaborate system of fixed fields. It is more likely that while the area immediately surrounding the settlements were more intensively cultivated, the outlying 'fields' were worked on a more temporary basis. It should be mentioned, however, that the presence of biennial weeds on some of the Polish sites may, in Wiślańsky's opinion, indicate the existence of a rudimentary form of the 2-field system.

Soil exhaustion often cited as the main reason for the abandonment of the settlements need not have been insurmountable problem in the Linear Pottery culture. The loess lands occupied by the Linear Pottery people provide one of the most fertile soils in Europe. Although the brown forest soils developed on loess do require some replenishment of nutrients and minerals to produce consistently high yields under continuous cultivation, this would not have been beyond the technological capability of the Linear Pottery farmers. Domesticated animals would have been pastured on the stubble, providing some manure for the plots. Household refuse may also have been used at least on the fields nearest to the settlements. The use of manure and kitchen refuse on the nearest fields is quite a common practice among shifting cultivators (Chisholm, 1968). Furthermore, legumes were known and cultivated by the Linear Pottery people and could have been used to restore depleted nitrogen to the soil.

Legumes are not particularly competitive with other protein foods in terms of nutritional value (Delwiche, 1978). Although they generally have a high protein content, they do not contain sufficient amounts of many of the essential amino acids. In addition, they are often difficult to digest, and a steady diet of pulses can lead to serious disorders, e.g., neurolathyrism and favism. The importance and survival of legumes as cultivated plants must have been, as Delwiche suggests, due, at least in part, to their ability to return nitrogen to the soil and thus help sustain the agricultural ecosystem.

Pollen evidence seems to indicate that a form of crop rotation may actually have been practiced at some of the Polish Linear Pottery sites (Wiślański, 1970). The purity of the grain finds and the presence of lentil seeds in the Szakálhát group would also suggest crop rotation. Before crop rotation, inter-cropping was probably used as a means of returning nitrogen to the soil. Manure and household refuse would have supplied the other essential nutrients as well as nitrogen.

Weed control may have been more of a problem in temperate Europe than in the Mediterranean. On the other hand, on the drier southern slopes competition between weeds and cultivated plants would have been less serious than on the wet bottom lands. At Pári-Altacker, for example, the grain cache did not contain a single weed grain.

It is possible that the increased amount of labour required for the weeding and manuring of the older plots may have been a factor in the abandonment of the settlements (cf. Carneiro, 1961). It should be emphasised however, that if the Danubian peasants did indeed practice shifting agriculture it would have been more out of choice than out of ecological necessity. When or where the establishment of permanent settlements became desirable or necessary, the farmers of the Linear Pottery culture seem to have been quite able to maintain such settlements. Population pressure toward the end of the Late Phase may have been influential in the creation of permanent settlements, at least in certain areas. The available evidence does not suggest that this stage was quite reached in Transdanubia although it seems to have been in some parts of the Alföld.

Nonetheless, even in Transdanubia, there is a noticeable increase in the amount of cultivated grains in the pollen diagrams of the Balaton toward the end of the Late Phase (Zólyomi, 1971). This, together with the archaeological evidence, would suggest an increase in agricultural production. The pollen evidence, even so, does not indicate intensive cereal or field agriculture until the Lengyel culture. One of the reasons for the difference in the development of the Alföld and Transdanubia may have been that the area of cultivable land was less circumscribed in the latter region (cf. Carneiro, 1961). In the Alföld, before the chernozem plateaus were brought into cultivation, cultivable land was restricted to those areas which were safe from flooding, but also had enough moisture to alleviate the vagaries of rainfall. In Transdanubia, on the other hand, rainfall was more reliable, while the loess lands provided large expanses of good quality, easily worked soils. When the need for farm products increased, it would have been ecologically possible to increase agricultural production by temporarily extending the arable, instead of resorting to more intensive methods. In historical times, Transdanubian peasants often added land to their more intensively cultivated plots by temporary clearings. The size of the clearance depended on the size and industry of the family (Kisvarga, 1965).

The change from the small house and nuclear family of the Alföld and southeast-Europe to the long house and extended family of the Linear Pottery culture of temperate Europe was probably a matter of difference in the organisation of labour. The example of the Kofyars of west Africa is interesting as it may have relevance to the European situation. 'Groups of formerly intensive cultivators on the plateau changed to shifting cultivation with longer fallows when the pressure of population density forced a budding-off onto the less densely occupied plains. The former pattern of dispersed homestead settlement, typical of the plateau hardly changed with the move, but in contrast there were significant changes in social organisation because of the new conditions of land-use. On the much larger bush farms of the plains a larger resident labour force, such as the extended family, offered a much higher production potential, so the plains households became larger with more polygynous marriages and greater frequency of extended families, unlike the situation on the plateau where the pattern of small co-resident households was the rule' (Netting, quoted by Smith, 1972, 417). The existence of the extended family in the Linear Pottery culture of Central and Western Europe has not, however, been sufficiently proven yet by archaeological evidence (Tringham, 1971). According to the evidence of the ground plan of the houses, there seems to have been a certain amount of regional variation in the internal construction of the Linear Pottery long houses. While the long house in eastern and central Europe may have housed an extended family, in western Europe it may only have been used by a nuclear family.

It is possible that in some areas part of the long house was used to shelter livestock. The younger and more valuable animals may have needed some protection against the colder winters of temperate Europe. Modderman (1970) has in fact suggested that the northwest end of the long house was used on the Dutch sites to stall animals. Although Modderman's suggestion is not backed by any direct evidence, it may be remembered in this context that the Linear Pottery settlements of the Netherlands are thought to have been occupied continuously without a break for several centuries. The manure from the stalls could have been used to sustain long-term cultivation.

The stalling of animals would have been one adjustment to the environmental conditions of midlatitude Europe. The change in stock composition, mentioned earlier, from sheep/goats to cattle and swine was another. It is likely that there were also differences in pasturing arrangements between temperate Europe and the Mediterranean. An important limitation in the former area was the shortage of winter fodder. Before the introduction of fodder crops, leaf fodder played a considerable part in the feeding of livestock (Clark, 1952). In the Mecsek region of Hungary, for example, oxen used daily for heavy transport overwintered until quite recently on little more than the young shoots and branches of oak trees (Andrásfalvy, 1970). Summer pasture, on the other hand, was more abundant, although not unlimited, than in the Mediterranean. On the loess plains of midlatitude Europe, transhumance would have been not only impractical but also largely unnecessary. The pasturing of the livestock could have been done by the individual families. In the forested regions of Transdanubia, until comparatively recently, it was the duty of the youngest and oldest male members of the family to pasture the family livestock (Andrásfalvy, 1972). Family pasturing in wooded areas was probably more efficient as it could make use of small glades and forest clearings. Moreover, the minding of larger herds would have been a more difficult task in the denser woodlands. No doubt, pasturing practices in the Linear Pottery culture varied a great deal depending on local conditions, as did probably the composition of the livestock. In suitable areas, transhumance may have been practiced on a limited scale, over short distances. But it is probably true that the connection between animal husbandry and cereal agriculture was more intimate in forested, temperate Europe than in the more arid lands of the Mediterranean.

SUMMARY AND CONCLUSIONS

A major change that occurred in Hungary during the spread of agriculture was the development of the Linear Pottery culture. The material evidence, demonstrating both continuity and change between the Körös and Linear Pottery cultures, has been described in detail by Tringham (1968, 1971) and Kalicz and Makkay (1972a; 1972b; in press). The aim of this study has been to show the continuity and changes that occurred in land use in the Körös and Linear Pottery transition.

The approach involved two basic procedures. First, to delineate the effective environments of the cultures the distributions of their settlements were analysed in relation to climate, vegetation, hydrography, geology and soils. Second, the environmental evidence thus obtained was correlated with the available archaeological evidence in a reconstruction of the economic base of the Körös, Alföld and Transdanubian Linear Pottery cultures.

Environmental Analysis

The results of the analysis, summarised in Appendix B, suggest that of the three cultures the environmental range of the Körös culture was the most limited. All the sites were located in the forest-steppe, in areas of insufficient moisture. 55% of the settlements are found on loess, and 40% on alluvial deposits. As to soils, 69% are on soil types in whose genesis water had been a decisive factor: alluvial, hydromorphic and alkali soils. At closer inspection, the types of environments occupied by the Körös culture are even more restricted. Of the 30% of sites that are found on chernozemic soils, most are in fact on meadow chernozems, which soil type is regarded as an intergrade between hydromorphic and chernozem soils and is often classed with the former (Szabolcs, 1961). The loess type occupied by the sites is mostly infusion loess, i.e., water-reworked loess. In many of its characteristics, it approximates alluvial deposits. The 80% figure indicating nearness to rivers underestimates reality. On a larger scale map, showing old oxbows and extinct small water courses, the figures would be much higher. All Körös sites were located immediately by rivers or lakes.

In contrast to the Körös culture, the range of environments occupied by the Alföld Linear Pottery culture is much wider. 29% of the sites are now in the forest zone; 21% on forest soils and 5% on chernozem-brown forest soil intergrades. Nonetheless, the similarities to the Körös culture are very marked. Only 7% of the settlements are located in areas of sufficient rainfall, and the largest percentage of sites, as shown in Appendix B, are in the same types of environments as the Körös sites.

The environmental range of the Transdanubian Linear Pottery culture is similar to that of the ALP. The percentage figures, however, indicate a definite shift to more humid forest environments: 66% of the sites are in the forest zone, and 50% in areas of sufficient climatic moisture. Correspondingly, there is an increase in the number of sites on forest soils: 34% on fertile forest soils, 27% on less fertile forest soils plus lithomorphic soils, and 7% on chernozem-brown forest intergrades. On larger scale soil maps, as indicated by the analysis carried out in the Marcal region, the percentage of sites on fertile brown forest soils and chernozem-brown intergrades would be considerably higher. As regards geological deposits, the highest percentage of sites (41%) are on loess, but most of the settlements are located on well drained parent materials. The majority of both ALP and TLP sites are found near rivers or lakes.

The percentage point difference, showing the joint effect of five categories of 14 variables on settlement distribution is relatively high for all three cultures: 66.74% for the Körös, 48.96% for the ALP, and 53.61% for the TLP. Some of the categories are,even so,very general, and it is safe to suggest that if they were further refined, the figures would be higher. In case of the ALP and TLP,the percentage point difference increases considerably when the calculations are made separately for the sites of each chronological phase, suggesting that changes occurred not only in the transition from the Körös culture but also during the life-span of the Linear Pottery culture.

The most noticeable are the changes that occurred during the last phase of both Linear Pottery cultures. In Late TLP, settlement shifts to drier areas: over 75% of the sites are in regions with insufficient moisture. This change is accompanied by a threefold increase in the number of sites located on chernozem-brown forest soils. A similar shift to drier areas has been observed by Quitta (1969) in some parts of Germany in the later phase of the Linear Pottery culture.

The late phase of the ALP is characterised by the development of regional groups, which have been defined mostly on pottery styles. The environmental analysis suggests that they also differed in the types of environments their settlements occupied. The Tiszadob group and the Bükk culture, of which the former is considered to have been the predecessor, are in fact very similar. They also differ from the rest of the groups. Perhaps, the least like the other groups is the Szakálhát, which, in its relationship to the environment, is strongly reminiscent of the Körös culture.

Changes in Land Use

The available evidence, both environmental and archaeological, suggests that the Körös culture was essentially a floodplain adaptation in a fairly dry climatic region. Rainfall in the Körös distribution area varies between 500-600 mm. However, evapotranspiration is very high, resulting in an annual water deficit of 100-175 mm. In addition, rainfall is very erratic in the Alföld. Partly to compensate for the lack of moisture, settlements were located immediately by rivers, mostly on levees. Along the Tisza, the sites today are found on the lower and upper Holocene terraces (Nagy, 1954). According to geomorphological studies, the present floodplain is the result of Subboreal downcutting (Somogyi, 1962), which would suggest that the sites in the Atlantic would have been lower in relation to the river, immediately in the floodplain.

Although the Körös people were farmers, who cultivated einkorn and emmer wheat, barley, millet and possibly flax and raised sheep/goats, cattle and pigs, the faunal evidence also indicates very heavy reliance on wild resources. In the three cultures, the percentages of wild animals is highest in the Körös culture. Most of these were animals of the rivers and the floodplain. Fish, shellfish, water birds, deer and wild swine from the gallery forests constituted an important, perhaps even a major part of the Körös diet. Although there is as yet little evidence for the direct utilisation of wild plants, it may be assumed that their share in the total diet would have been similarly high.

According to the evidence of avifaunal remains, the settlements were occupied all the year round (Bökönyi, 1972). In contrast to the ALP and TLP sites, the cultural deposits found at the Körös sites are relatively thick, suggesting that occupation was of relatively long duration. Although there are no Körös tells, according to Kalicz (1965), the quantity of material remains is reminiscent of tell-like settlements.

In the relatively permanent nature of the settlements, the richness of the river and floodplain resources undoubtedly played a decisive role. Agriculture, in comparison, seems to have been less important. Even so, the permanence of the settlements suggests that agriculture could be maintained in the same place for a considerable period of time. Levee soils, mostly chernozems and meadow chernozems today, are among the most fertile soils. Certain amounts of fertilizers could also be provided by animals grazing on the stubble or by the addition of kitchen refuse.

The levees also offered a number of other advantages for crop cultivation. They were well drained and at the same time, during the floods, usually the critical period in the growing season, the rising ground water table would have been within reach of the crops. The levees were coarser textured than the surrounding floodplain and provided lighter soils, which were easier to work. Being near the rivers, their groundwater was less saline (Várallyay, pers. comm.). Due to their higher position, they offered some degree of protection against frosts in winter. They were also relatively safe from flooding, which last was an important consideration in the Alföld before the regulation of the rivers. The Danube and the Tisza and their tributaries have at least two regular annual floods: one in early spring, the other early summer. Whilst in the Atlantic, before man's removal of much of the vegetation cover in the upper reaches of the rivers, the vehemence of the floods may have been less, the danger was nonetheless real. Some of the settlements were abandoned because of flood destruction.

The seasonality and vehemence of the floods would have been important in influencing decisions on crop policy. It is unlikely that the Körös farmers would have relied on a single crop or even a single area of cropping. Einkorn, which is poorly suited to irrigation agriculture (Butzer, 1965) was probably mostly grown on the levee tops and slopes. Emmer and barley may have been grown in both low and high lying ground. In case of floods and other disasters, millet could be sown in the floodplain after the floodwaters had receded. Millet has a short growing season,and in the recent past it was often used to supplement the main crop (Kiss, 1929). In Körös times, it could also have reduced the risk of total crop failure.

Among the domesticated animals, sheep in the Körös culture were predominant. Cattle and pigs seem to have been less important. In the predominance of sheep, beside cultural tradition, as suggested by Bökönyi (1971), the presence of alkaline pastures near the settlements may also have been a factor. Both paleobotanists and pedologists suggest the existence of some salt affected soils in the Alföld in the Atlantic. The relatively permanent nature of the fields may also have contributed to the numerical importance of sheep (cf. Clark, 1947).

Beside their other uses, the floodplains were also important in providing pasturage for the domesticated stock. Before the introduction of fodder crops, the scarcity of winter fodder was a major limiting factor in stockbreeding in temperate Europe. According to ethnohistorical evidence, the floodplains and marshes of the Alföld had, before the regulation of the rivers, provided adequate fodder for sizeable herds in winter (Glaser, 1939). In addition, the floodplains could also be used for summer grazing.

Until recently, livestock in the Alföld were mostly kept in the open all the year round. Until the end of April, sheep were kept near the settlements, mostly in the fields, to thin out the crop. In summer until mid-July, they were usually grazed on the higher loess plateaus; after July, when the upper pastures began to feel the effect of drought, the flocks were transferred to the floodplains. In autumn, sheep were again kept near the villages, mostly on the stubble. Winter fodder was provided by the floodplains. It seems likely that the Körös farmers would have followed a similar schedule.

Archaeological and botanical evidence can provide some indication of the scheduling of some of the other subsistence activities. On the basis of a pit excavated at Röszke-Luvár, it has been suggested that fishing was probably practiced in summer, the hunting of forest animals in autumn and winter, and the gathering of shellfish in spring (Trogmayer, 1968). As regards cereal cultivation, the presence of a Bromus species among the weeds at Gyálarét-Szilágyi major might indicate that at least some of the wheat was autumn sown. Millet and flax would have been sown in late spring-early summer.

The available evidence indicates little trade or exchange between the sites in the Körös culture. The settlements were probably largely self-sufficient, except for a limited amount of raw materials, such as stone, which must have come from outside the Alföld. In comparison with the ALP, Körös settlements are large although there is naturally some variation in size. As Kalicz (1957) has observed, the sites are more or less evenly spaced. On the whole, there seems to have been no development toward a hierarchy of settlements. While it would be difficult to give an estimate of the Körös population, the lack of tell sites does suggest a certain mobility in settlement location and would indicate that competition for land could not have been too severe. Rivers in the Alföld often changed their beds, which may have been an additional reason for the frequent abandonment of the settlements.

In the Linear Pottery culture, the wheat-sheep-aquatic/wild resource economic pattern of the Körös culture underwent several changes, the best documented of which is the shift to locally available domesticated animals. Cattle replaced sheep as the most important species, and pigs, which had been almost negligible at the Körös sites, became equally or more important than sheep. Aurochs and wild swine are both indigenous to much of temperate Europe,and the increase in the number of their domesticated counterparts at the Linear Pottery sites was without doubt the result of local domestication.

The shift to cattle and pigs meant not only a larger pool of wild animals available for domestication (Bökönyi, 1974), but also more meat supplied by the domesticated stock. Cattle needs 5 times as much pasturage than sheep (Györffy and Visky, no date), but provides 10 times as much usable meat (Flannery, 1971). The pasturage requirements of pigs are roughly equal to those of sheep, while the usable meat from pigs is 3 times greater. Faunal evidence suggests that hunting in the Linear Pottery culture was less important than in the Körös culture; of the early and middle neolithic cultures of Hungary, the Late TLP has the smallest percentage of wild animals. Evidence from the ALP also indicates some decline in the importance of hunted animals. Alongside with hunting, fishing and the exploitation of other aquatic resources also decreased in both Linear Pottery cultures.

The evidently successful domestication of pigs and cattle in addition to the caprovines meant greater flexibility and allowed for greater diversity in the composition of the domesticated stock. Although cattle was invariably the most frequent domesticated animal, the ratio of sheep and pigs did vary from settlement to settlement. Greater diversity between settlements opened up the possibility of at least some degree of local specialisation and of interchange of food produce.

A further consequence of the emphasis on domesticated cattle was that it allowed early farming man to exploit the forest-steppes more productively. At the technological level of the Körös and Linear Pottery cultures, the high plains offered little suitable land for cereal agriculture. Although the farmers of the Körös culture had exploited the steppes to some extent through hunting and the pasturing of sheep, the extent of exploitation was fairly limited. Through his domesticated cattle, the forest steppes began to provide man with a substantial part of his meat supply.

Although cattle is relatively efficient in terms of meat production/unit land area, it does require very extensive pasturage. The break up of the larger Körös sites and the appearance of the numerous but very small ALP settlement probably supporting not more than 1 or 2 families may have been connected with the change to cattle domestication and the importance of stockbreeding. In forested Transdanubia, where both summer and winter pasture were more limited, cattle breeding was probably less extensive, even if cattle was still the most important domesticated animal. On the other hand, the attraction of the forest-steppe as potential pasture may have been a consideration in the occupation of the drier margins by the TLP settlers toward the end of the late phase.

In plant cultivation, there seems to have been no similar shift to locally available species, although rye may already have been domesticated at some of the Polish Linear Pottery sites (Wiślański, 1970). The evidence does, however, suggest the cultivation of a wider range of domesticated plants, cereals and legumes, in the Linear Pottery culture than in the Körös culture. In addition to einkorn and emmer wheat, two-row hulled barley and millet, the Linear Pottery farmers cultivated breadwheat, six-row hulled barley and lentil.

Millet in wooded Transdanubia was probably sown as the first crop after forest clearance. The presence of Agrostemma githago along the weeds at Zánka indicates that at least some of the wheat was autumn sown (cf. Tempir, 1964).

Six-row hulled barley is first identified in Hungary in the Szakálhát group, where it occurs in some quantity. The separately stored barley and wheat finds at Dévaványa-Réheli gát would suggest separate sowing and harvesting for each crop (Hartyányi and Nováki, in press). The finds, which are remarkably pure, would argue for good weed control.

The agricultural developments of the Szakálhát group closely parallel developments that occurred further south. Greater diversity in the exploitation of cereal crops, the growing importance of barley, and crop purity were also characteristic features of late neolithic agriculture in Greece. As Renfrew (1972) has suggested, increasing crop purity opened up new possibilities for further sophisticated advances, such as the rotation of crops. The Szakálhát people did in fact cultivate lentil, which could have been used in crop rotation.

Renfrew has also suggested that the plough may have been known in late neolithic/chalcolithic Greece. According to Ghetie and Mateescu (1973), certain changes in the thoracic vertebrae and long bones of some bovids found at the sites of Vădastra and Crusovu would indicate that cattle had been yoked and used for traction in the Vădastra culture. The enormous quantities of wheat that were found on the sites would be consistent with the use of the plough. The location of the Szakálhát settlements near larger tracts of fertile meadow chernozems might also be an indication of the use of the plough. Nonetheless, direct evidence that the plough was actually known in Europe at this time is so far lacking.

The appearance of stratified settlements together with the botanical evidence suggests that agriculture was becoming more intensive in the Alföld during the late phase of the Linear Pottery culture. There is no similar indication of more intensive land use from Transdanubia before the advent of the Lengyel culture, although the growing amount of cultivated grains in the pollen diagrams of the Balaton (Zólyomi,, 1971) and the large number of querns found,for example,at Sukoró-Tóradülö would seem to suggest increasing reliance on cultivated crops toward the end of the Linear Pottery culture.

One reason for the difference in land use between the Alföld and Transdanubia may have been that good agricultural land was more abundant in the latter region (cf. Carneiro, 1961). In the drier climate of the Alföld, cereal cultivation was mostly limited to the floodplains, where suitable land was greatly circumscribed. In wooded Transdanubia, where moisture was less limiting and where the forested loess lands provided large expanses of easily worked, well drained soils, cereal agriculture was no longer restricted to the gallery forests; the major resource area, though the floodplains and marshes would still have been exploited, shifted to the zonal forests. When the need for farm produce increased, it would have been possible to increase agricultural production by temporarily extending the arable, instead of resorting to more intensive methods.

In most systems of shifting cultivation, the fields immediately by the settlement, usually located on the most fertile part of the soil catena, are cultivated continuously (Chisholm, 1968). Soil fertility is maintained by animal manure, where available, or by the addition of kitchen refuse. The location of the TLP sites on the most productive soil and loess types of the area might indicate a similar system of continuously cultivated plots by the settlements. The less fertile interfluves, covered by lessivated or podzolised brown forest soils, could have been cultivated on a temporary basis.

Land exhaustion, often cited as the main reason for the shifting pattern of settlement and cultivation, need not have been a limiting factor in the Linear Pottery culture. Other considerations, among which increased labour requirements needed for weeding or maintaining the plots may have played a part, were probably more decisive. While land was abundant, shifting cultivation was a possible and also efficient form of land utilisation (Carneiro, 1961; Conklin, 1962). However, it was not an environmentally imposed necessity. While shifting cultivation may have been the general rule, permanent settlements and, by implication, more permanent agriculture also existed in the Linear Pottery culture. Some of the Dutch sites were occupied continuously for some hundred years (Moddermann, 1970). In Poland, on the basis of botanical evidence, Wiślański (1970) has suggested the possibility that a form of two field rotation was practiced.

The extended family implied by the appearance of the long house in the Linear Pottery culture may have been a response to the different labour requirements of the more extensive forest agriculture of midlatitude Europe. Changes from the nuclear to the extended family as a consequence of changes in land use are known to have occurred in other parts of the world (Smith, 1972). The contrast between the small house of the Alföld and the long house of Transdanubia is noteworthy in this respect although small, often round houses are also found in the latter area.

More intensive agriculture in the Alföld in the late phase was paralleled by more intensive exploitation of other raw materials, especially obsidian. While obsidian had been present at the Körös sites, it always constituted a small percentage of the total lithic assemblage. The dramatic increase in the quantity of obsidian tools at the Szatmár sites would suggest direct exploitation at the source by the Linear Pottery people: most early sites are located near or en route to the obsidian deposits of the Tokaj Mountains. Exchange or trade was, however, mostly local; obsidian in Vinča A is still relatively rare. It was in Vinča B, which parallels the Late phase, that obsidian becomes numerous for the first time. The presence of obsidian together with Tiszadob and Bükk sherds as far north as Poland and with Szakálhát sherds as far south as Yugoslavia and even Thessaly would indicate a considerably expanded exchange network, both in volume and distance in the Late phase. Increased production is further suggested by the appearance of specialised settlements in the Bükk culture.

The expansion of long distance trade was probably strongly influenced by developments, similar to those of the Szakálhát group, that occured in the Vinča culture and elsewhere in southeast Europe at this time. The settlement of the lands to the north also helped increase demand.

Although long distance trade was probably mostly concerned with precious and durable items such as obsidian and Spondylus shells, local trade or exchange may also have involved some food produce; sherds of one regional group are commonly found at the sites of another in the Alföld. Certain regularities observed in the spacing and size of the Szakálhát settlements (Kalicz, 1957) could be interpreted as reflecting the emergence of an incipient form of settlement hierarchy in the Alföld. Although this hypothesis is not yet confirmed by archaeological evidence, in view of the other developments that took place in the Late phase some differentiation in the function and organisation of the settlements seems not unlikely.

While the multiplication of settlement units and increased agricultural production clearly indicate an expanding population in both the Alföld and Transdanubia in the Linear Pottery culture, the appearance of stratified sites, hilltop settlements and more intensive land use in the Szakálhát and Bükk groups would suggest greater competition for resources and probably a higher population density in relation to usable agricultural land in the Alföld. In Transdanubia, although there were close links with eastern Hungary and Vinča, the evidence so far does not show a similar development toward a hierarchy of settlements. It should be pointed out, however, that the available evidence is very limited, and the chronological position of many of the Transdanubian sites in relation to the Notenkopf and Zseliz groups is somewhat uncertain. The settlement of the drier forest steppe margins in the Late TLP would imply not only an expanding population but also at least some measure of competition for land at the close of the Linear Pottery culture.

Man's Effect on the Environment

According to the pollen evidence, the immigration of Mediterranean species in Hungary reached its maximum in the Atlantic phase (Zólyomi, 1958). While alterations in climatic conditions were probably important or even decisive, some of the changes were undoubtedly anthropogenic. A number of new species, cereals, legumes and sheep and goats, were introduced intentionally by farming man, and the introduction of some other species, weeds, parasites and pests, though unintentional, was also due to man. Plant cultivation, the grazing of animals and improved communication between settlements also helped alter the old species composition.

The periodic increases in the proportion of birch, pine and Pteridiophyta pollen observed in the diagrams of the Nyirség region, which have been interpreted as climatic oscillations (Csinády, 1959; 1959-60), may in fact have been anthropogenic. The fluctuations could have been occasioned by slash-and-burn agriculture; birch and pine are usually the first trees to appear after a forest fire (Iversen, 1956). The Nyirség was well settled by the Linear Pottery culture, and the cultivated cereals which first appear in the pollen diagrams at this time are positive evidence for the presence of farming man in the area.

Nonetheless, man's ability to change the landscape in the ecologically less fragile environments of temperate Europe was itself limited. At the technological level of the early farming communities, plant cultivation and stockbreeding, limited by a long dormant season, winter cold, disease and pests, must have been relatively small-scale and often transient. While it must be admitted that relatively little is known about the effect that early farming man in Hungary had on his environment, what evidence there is, does not indicate widespread ecological disturbance. Neither on the periphery of the Alföld, nor in Transdanubia, is there any evidence according to the pollen diagrams of large scale permanent deforestation during the period contemporary with the Körös and Linear Pottery cultures.

As regards soils, man may have been influential in the transformation of some brown forest soils into forest-chernozem intergrades. These soils are located in climatically transitional areas and thus man's influence may have been more decisive. Edaphic conditions would have delayed reforestation once the original forest cover had been destroyed. But again it is unlikely that this process could have been completed during the neolithic. In the creation of chernozems along the levees, man's removal of the original vegetation cover may also have been a complementary factor. Similarly, in the transformation of some brown forest soils into rendzinas grazing probably played a part (Góczán, 1971). Erosion, however, as much of the forest cover was still intact, was probably not yet a serious problem in the early and middle neolithic, although some loss of top soil would have occurred with hillslope cultivation.

In Hungary, at least, it is unlikely that man was directly responsible for the creation of chernozem soils which occupy the higher loess plateaus. Buried chernozems have been found under Bronze Age barrows in the Alföld (Borsy, 1966-67), one of which has been dated to 2370 bc. According to Borsy (pers. comm.), the full development of these soils required a minimum of 1500 but more likely 2-3000 years. The chernozems were probably formed in the Boreal period, when the pollen evidence shows extensive steppe vegetation in Hungary. Although the Atlantic phase was wetter, typical forest-steppe and steppe vegetation, according to the pollen evidence, still survived in the Alföld (Járai-Komlódi, 1966). The existence of chernozem soils during the Atlantic is indirectly confirmed by the lack of settlement on the high plateaus. As Ložek (1967) has suggested, man's role was more to protect the already existing steppe and forest-steppe enclaves from the invasion of the forest than to bring about their existence through direct action.

Conclusions

Although it was possible to identify a few of the changes that occurred in the land use patterns of the earliest farming cultures of Hungary, this study of Körös and Linear Pottery environmental relationships is far from complete. The reconstructions presented in the foregoing part are tentative. Floral and faunal data showing how the environment was exploited are only available for a small percentage of the sites. Pollen evidence on the basis of which man's effect on the environment could be assessed is still limited. In attempting to determine the environmental relationships of the Körös and Linear Pottery cultures certain generalisations have been unavoidable: local variations have been to a large extent unexplored. More detailed studies could uncover relationships which on the smaller scale maps went undetected. Finally, to evaluate the Hungarian results we need to know more about the changes that occurred outside Hungary. At present, similarities and differences with other areas of the Karanovo-Starčevo-Criş and Linear Pottery cultures are only vaguely perceived; the problem of regional and temporal variations that must have occurred in these cultures awaits further study.

REFERENCES

Andó, M. and Vágás, I., 1972. A Tiszavölgy 1970. évi nagy árvize. Földr. Közl. XX, 18-36.

Andrásfalvy, B., 1970. A mohácsiak állattartása 1686-tól 1848-ig. II. Tanulmányok a Dél-Dunántul Történetéből II (1969), 153-191.

Andrásfalvy, B., 1972. Dunántuli pásztorkodás - alföldi pásztorkodás, in Tájkutatási Tudományos Ülésének Előadásai (1970), MTA Dunántuli Tudomány Intézete, Pécs.

Bacsó, N., 1971. The climate of Hungary, in Pécsi, M., Enyedi, Gy. and Marosi, S. (Eds.), Hungary – Geographical Studies, Budapest, 65-83.

Bakay, K., 1970. Nagytarcsa, Mitt. des Arch. Inst. der Ungarischen Akad. der Wissenschaften 1. Zehn Jahre Archäologische Forschung (1958-1968), 121.

Bakay, K., 1971. Szeghalom, Kovácshalom. Mitt. des Arch. Inst. der Ungarischen Akad. der Wissenschaften 2. Archäologische Forschungen 1969, 135-136.

Bakay, K. et al., 1966. Veszprém megye régészeti topográfiája. A keszthelyi és tapolcai járás. Magyarország Régészeti Topográfiája 1, Budapest.

Bakay, K., et al., 1970. Veszprém megye régészeti topográfiája. A deveceseri és sümegi járás. Magyarország Régészeti Topográfiája 3, Budapest.

Bálint, A. and Szabó, J., 1955. Leletmentő ásatások as 1953. évben. Arch. Ért. LXXXII, 95.

Balogh, K. et al., 1958. Magyarázó Magyarország 1:300 000-es földtani térképéhez. Magyaro. Áll. Földtani Int. Budapest.

Bándi, G., 1969. Az 1968. év régészeti kutatásai. Rég. Füz. 22, 21.

Bánki, Zs., 1966. Az 1965. év régészeti kutatásai. Rég. Füz., 19, 7.

Bánki, Zs., 1967-68. Régészeti kutatások. Alba Regia VIII-IX, 179-180.

Bánki, Zs., 1969. Régészeti kutatások. Alba Regia X, 149-151.

Bánki, Zs., 1970. Régészeti kutatások. Alba Regia XI, 163-166.

Banner, J., 1928. Az ószentiváni ásatások. Dolgozatok Szeged IV, 148-237.

Banner, J., 1931. A neolithikum Szarvason. Dolgozatok Szeged VII, 61-67.

Banner, J., 1931a. Agyagból készült láda a neolithikumból. Arch. Ért. XLV, 253-256.

Banner, J., 1932. A kopáncsi és kotacparti neolithikus telepek és a tiszai kultura III. periodusa. Dolgozatpk Szeged VIII, 1-32.

Banner, J., 1935. Ásatás a hódmezővásárhelyi Kotacparton. *Dolgozatok Szeged* XI, 97-120.

Banner, J., 1937. Die Ausgrabungen des Museums vom Hódmezővásárhely im Jahre 1935. *Dolgozatok Szeged* XIII, 70-77.

Banner, J., 1937a. Die Ethnologie der Körös Kultur. *Dolgozatok Szeged* XIII, 32-49.

Banner, J., 1938. A hódmezővásárhelyi városi múzeum ásatásai 1937-ben. *Dolgozatok Szeged* XIV, 191-198.

Banner, J., 1940. *Hódmezővásárhely Története a Honfoglalás Koráig* I, Hódmezővásárhely.

Banner, J., 1942. *Das Tisza-, Maros-, Körös Gebiet bis zur Entwicklung der Bronzezeit.* Szeged.

Banner, J., 1942a. Ujabb ásatás a hódmezővásárhelyi Kökénydombon. *Dolgozatok Szeged* XVIII, 138-140.

Banner, J., 1943. Az ujabbkökori lakóházkutatás mai állása Magyarországon. *Arch. Ért.* LXX. 1-25.

Banner, J., 1951. A harmadik ásatás a hódmezővásárhelyi Kökénydombon. *Arch. Ért.* LXXVIII, 27-36 .

Banner, J. and Bálint, A., 1935. A szakálháti őskori telep. *Dolgozatok Szeged* XI, 76-88.

Banner, J., and Foltiny, I., 1945. Ujabb ásatások a hódmezővásárhelyi Kökénydombon. *Fol. Arch.* V, 8-34.

Banner, J. and Korek, J., 1949. Negyedik és ötödik ásatás a hódmezővásárhelyi Kökénydombon. *Arch. Ért.* LXXVI, 9-25.

Banner, J., and Párducz, M., 1948. Ujabb adatok Dél-Magyarország ujabb kőkorához. *Arch. Ért.* LXXIX, 19-41.

Banner, J., Bóna, I. and Márton, L., 1959. Die Ausgrabungen von L. Márton in Toszeg. *Acta Arch. Hung.* X, 1-140.

Buckman, H.O. and Brady, N.C., 1969. *The Nature and Properties of Soils.* New York.

Buol, S.W., Hole, F.D. and McCracken, R.J., 1973. *Soil Genesis and Classification.* Ames.

Buringh, P., 1957. Living conditions in the Lower Mesopotamian Plain in Ancient Times. *Sumer* 13, 30-46.

Butzer, K.W., 1961. Climatic change in arid regions since the Pliocene, in Stamp, D. (ed.) *A History of Land Use in Arid Regions* UNESCO, 31-56.

Butzer, K.W., 1965. Physical Conditions in Eastern Europe, Western Asia and Egypt before the period of agricultural and urban settlement. *Cambridge Ancient History*, I.

Butzer, K.W., 1971. *Environment and Archaeology–An Ecological Approach to Prehistory.* London.

Carneiro, R.L., 1961. Slash-and-burn cultivation among the Kuikuru and its implications for cultural development in the Amazon basin, in Wilbert, J., (Ed.) The Evolution of Horticultural Systems in Native South America. Causes and Consequences. A Symposium. Caracas, 47-67.

Castiglione, L. and Salamon, E. (Eds.) 1970. Zehn Jahre Archäologische Forschung (1958-68). Mitt. des Arch. Inst. der Ungarischen Akad. der Wissenschaften 1.

Centre d'Etude de Azote, 1960. Progressive Wheat Production. No.4, Geneva.

Childe, V.G., 1929. The Danube in Prehistory. London.

Childe, V.G., 1957. The Dawn of European Civilisation. New York.

Chisholm, M., 1968. Rural Settlement and Land Use–An Essay in Location. London.

Clark, J.G.D., 1947. Sheep and swine in the husbandry of Prehistoric Europe. Antiquity XXI, 122-136.

Clark, J.G.D., 1952. Prehistoric Europe–The Economic Basis. London.

Clark, J.G.D., 1965. Radiocarbon dating and the expansion of farming culture from the Near East over Europe. Proc. Prehist. Soc. XXXI, 58-73.

Clark, G., and Piggott, S., 1965. Prehistoric Societies. London.

Clarke, D.L., 1968. Analytical Archaeology. London.

Clarke, D.L., 1976. Mesolithic Europe: the economic basis, in Sieveking, G. de G., Longworth, I.H., and Wilson, K.E. (Eds.) Problems in Economic and Social Archaeology, London, 449-481.

Coe, M.D. and Flannery, K.V., 1971. Microenvironments and Mesoamerican Prehistory, in Struever, S. (Ed.) Prehistoric Agriculture, New York, 131-142.

Conklin, H.C., 1962. An ethnoecological approach to shifting cultivation, in Wagner, Ph.L. and Mikesell, M.W. (Eds.) Readings in Cultural Geography, Chicago, 457-464.

Csallány, D., 1958. Az 1957, év régészeti kutatásai. Rég. Füz. 10, 39.

Csallány, D., and Gombás, A., 1963. Az 1962. év régészeti kutatásai. Rég. Füz. 16, 45.

Csallány, D., and Gombás, A., 1965. Az 1964. év régészeti kutatásai. Rég. Füz. 18, 6.

Csallány, G., 1939. Gesichtsdarstellung auf Gefassen der Theisskultur. Germania, 23, 145-46.

Csalog, J., 1941. A magyarországi ujabb-kőkori agyagművesség bükki és tiszai csoportja. Fol. Arch. III-IV, 1-27.

Csalog, J., 1941a. A Vucedol-Zóki tipusú kerámia lelőhelyei Tolna vármegyében. Arch. Ért. 68, 6-14.

Csalog, J., 1941b. Középkori falusi kút Medina határában. Dolgozatok Szeged XVII, 173-75.

Csalog, J., 1942. Bronzkori temető és ujabb-kőkori lakótelep nyomok Bonyhád határában. Arch. Ert. 69, 119-43.

Csalog, J., 1955. A Tiszai müveltség viszonya a szomszédos újkőkori müveltségekhez. Fol. Arch. VII, 23-44.

Csalog, J., 1956. Ujkőkori gyékénylenyomat Kökénydombról. Arch. Ért. 83, 183-85.

Csalog, J., 1958. Az 1956. év régészeti kutatásai. Rég. Füz. 9, 10-13.

Csalog, J., 1959. Az 1958. év régészeti kutatásai. Rég. Füz. 11, 21-22.

Csalog, J., 1960. Az 1959. év régészeti kutatásai. Rég. Füz. 13, 29.

Csalog, J., 1964. Az 1963. év régészeti kutatásai. Rég. Füz. 17, 17-18.

Csalog, J., 1965. Az 1964. év régészeti kutatásai. Rég. Füz. 18, 20.

Csalog, J., 1966. A legújabb kökénydombi fonatlenyomat tanulságai. A Móra F. Múzeum Évkönyve (1964-65), 17-45.

Csalog, Zs., 1961. Az 1960. év régészeti kutatásai. Rég. Füz. 14, 20.

Csalog, Zs., 1964. Az 1963. év régészeti kutatásai. Rég. Füz. 17, 48.

Csalog, Zs., 1965. Az 1964. év régészeti kutatásai. Rég. Füz. 18, 4.

Csalogovits, J., 1930. Földrajzi tényezők hatása Magyarország neolitikus kulturáinak kialakulására és elterjedésére. Arch. Ért. XLIV, 28-52.

Csánk, G.V., 1964. Megfigyelések a békásmegyeri őskori telepen. Arch. Ert. 91, 201-214.

Csánk, G.V., 1967. Az 1966. év régészeti kutatásai. Rég. Füz. 20, 9.

Csinády, G., 1954. A Bátorligeti láp története a pollenanalizis tükrében. Földr. Ért. III, 684-689.

Csinády, G., 1959. A Csarodai láposodott folyómeder pollenanalitikai vizsgálata. Acta Univ. Debreceniensis V (1958), 271-77.

Csinády, G., 1959-60. A Kokadi láp palynologiai vizsgálata. Acta Univ. Debreceniensis VI, 239-51.

Czógler, K., 1933-34. Édesvizi kagylók szegedvidéki régészeti leletekben. Dolgozatok Szeged IX-X, 298-302.

Daniel, G.E., 1950. A Hundred Years of Archaeology. London.

Dankó, I., 1964. A bodrogközi Hosszú-rét települése. A Herman O. Múzeum Évkönyve (1962-63) Iv, 143-63.

Darnay, K., 1905. Szeghalmi ásatásokról. Arch. Ért. XXV, 66-70.

Dax, M. 1967. Az 1966. év régészeti kutatásai. Rég. Füz. 20, 22.

Dax, M. et al., 1972. Veszprém megye régészeti topográfiája. A pápai és zirci járás. Magyarország Régészeti Topográfiája 4, Budapest.

Delwiche, C. C., 1978. Legumes–Past, Present and Future. BioScience, Vol. 28, No.9, 565-570.

Dombay, J., 1958. Kőrézkori és kora-vaskori település nyomai a Pécsváradi Aranyhegyen. A Jan. Pann. Múzeum Evkönyve, Pécs, 53-96.

Dornyay, D., and Csalog, J., 1945. Az újabb-kőkori szalagdiszes agyagműveltség emlékei a Balatoni Múzeumban. Fol. Arch. V, 1-7.

Draveczky, B., 1966. Neuere Angaben zur Verbreitung der Linearkeramik im südlichen Teil von Transdanubien. Acta Antiqua et Arch. X, 27-33.

Draveczky, B., 1970. Somogy megye régészeti képeskönyve. Somogyi Múzeum, Kaposvár, no. 17.

Draveczky, B. and Garam, E., 1967. Az 1966. év régészeti kutatásai. Rég. Füz. 20, 52.

Draveczky, B., and Sági, K., 1964. A somogy megyei múzeumok régészeti adattára. Somogyi Múzeum, Kaposvár, No. 2.

Ébner, S., 1925. A Bodrogköz lápi községeinek település földrajzi vázlata. Föld és Ember V, 65-102.

Ecsedy, I., 1971. Az 1970. év régészeti kutatásai. Rég. Füz. 24, 5-6.

Ecsedy, I., 1972. Neolitische Siedlung in Dévaványa-Katonaföldek. Dévaványa-Réhelyi gát. Mitt. des Arch. Inst. der Ungarischen Akad. der Wissenschaften 3, 59-63, 153-154.

Ecsedy, I. 1972a. Az 1971. év régészeti kutatásai. Rég. Füz. 25, 20.

Edelmann, C. H., 1950. Some unusual aspects of soil science. Fourth Int. Cong. Soil Science, Amsterdam. 17-25.

Enyedi, Gy., and Szabó, G. M., 1955. Öntözéses gazdálkodás a Délkelet-Alföldön. Földr. Közl. III, 333-347.

Erdélyi, I., 1965. Az 1964. év régészeti kutatásai. Rég. Füz. 18, 16.

Éri, I., 1962. Az 1961. év régészeti kutatásai. Rég. Füz. 15, 74.

Éri, I. et al., 1969. Veszprém megye régészeti topográfiája. A veszprémi járás. Magyarország Régészeti Topografiája 2, Budapest.

Éri, K. and Kőszegi, I., 1957. Leletmentő ásatások az 1955. évben. Arch. Ert. 84, 85.

Eyre, S. R., 1963. Vegetation and Soils–A World Picture. London.

Fitz, J., 1963-64. Jelentés a Fejér megyei múzeumok 1960-62. évi tevékenységéről. Alba Regia IV-V, 172-76.

Firbas, F., 1949-52. Spät- und Nacheiszeitliche Waldgeschichte Mitteleuropas Nördlich der Alpen. 2 Vols. Jena.

Flannery, K. V., 1971. Origins and ecological effects of early domestication in Iran and the Near East, in Struever, S. (Ed.) Prehistoric Agriculture, 50-79.

Flannery, K.V., 1972. The origins of the village as a settlement type in Mesoamerica and the Near East: A comparative study, in Ucko, P., Tringham, R. and Dimbleby, G.W. (Eds.) Man, Settlement and Urbanism, London, 23-53.

Flannery, K.V. et al., 1971. Farming systems and political growth in Ancient Oaxaca, in Struever, S. (Ed.) Prehistoric Agriculture, 157-78.

Foltiny, I., 1941. Korarèz ès Bronzkori temetö Deszken. Fol. Arch. III-IV, 69-88.

Ford, R.I., 1972. An ecological perspective on the eastern pueblos, in Ortiz, A. (Ed.) New Perspective on the Pueblos. Albuquerque.

Frenzel, B., 1966. Climatic change in the Atlantic/sub-Boreal transition on the Northern hemisphere: Botanical evidence. World Climate from 8000 to 0 B.C. Royal Met. Soc. London, 99-123.

Gállos, F., 1960. Vázlatok Pécsvárad kialakulásának és középkori mezővárosi fejlődésének történetéhez. A Jan. Pann. Múzeum Évkönyve Pécs, 159-180.

Gallus, S., 1936. A nagytétényi neolitikus sir. Arch. Ert. XLIX, 85-86.

Galván, I., 1958. Az 1957. év régészeti kutatásai. Rég. Füz. 10, 19.

Garasanin, D., 1961. Die Siedlung der Starčevo kultur in Nosa bei Subotica und das Problem der neolitischen Lehmschermen. Fifth Congress ISPP Hamburg 1956, 303-307.

Garnett, A., 1945. The loess regions of Central Europe in prehistoric times. Geogr. Journal 106, 132-143.

Gazdapusztai, Gy., 1957. A Körös kultura lakótelepe Hódmezővásárhely-Gorzsán. Arch. Ért. 84, 3-13.

Gazdapusztai, Gy., 1959. Az 1958. év régészeti kutatásai. Rég. Füz. 11, 16-17.

Gazdapusztai, Gy., 1964. Az 1963. év régészeti kutatásai. Rég. Füz. 17, 14.

Gazdapusztai, Gy., 1966. Az 1965. év régészeti kutatásai. Rég. Füz. 19, 9.

Gazdapusztai, Gy., 1967. Az 1966. év régészeti kutatásai. Rég. Füz. 20, 15.

Gazdapusztai, Gy., and Draveczky, B., 1967. Az 1966. év régészeti kutatásai. Rég. Füz. 20, 14.

Ghetie, B.C. and Mateescu, C.N. 1973. L'utilisation des bovins à la traction dans la Neolithique Moyen (D'après les nouvelles observations osteologiques faites dans les sites de Vádastra et de Crusovu, Roumanie.) Eight Congres ISPP Beograd 1971, 454-461.

Glaser, L., 1939. Az Alföld régi vizrajza és a települések. Földr. Közl. LXVII, 297-307.

Gnandt, K.P., 1957. Leletmentő ásatások az 1955. évben. Arch. Ért. 84, 85.

Góczán, L., 1971. A Marcal-Medence Talajföldrajza. Földrajzi Tanulmányok 12, Budapest.

Góczán, L., 1972. Soil geography of the Marcal basin. MTA Geographical Research Inst. Abstracts No. 17.

Gombás, A., 1960. Juhtartás Szentmihályon (Tiszavasvári). A Nyiregyházai Jósa A. Múz. Évkönyve III, 223-245.

Gradmann, R., 1906. Beziehungen zwischen Pflanzengeographie und Siedlungsgeschichte. Geogr. Zeit. 12, 305-25.

Gradmann, R., 1933. Die Steppenheide theorie. Geogr. Zeit. 39, 265-78.

Green, E.L., 1973. Location analysis of prehistoric Maya sites in Northern British Honduras. American Antiquity 38, 279-293.

Grigg, D., 1967. Regions, models and classes, in Chorley, R.J. and Haggett, P. (Eds.) Integrated Models in Geography, Parts I and IV, London.

Györffy, J. and Visky, K., no date, A Magyarság néprajza. Vol. II. Gazdálkodás. Budapest.

Haggett, P., 1965. Locational Analysis in Human Geography. London.

Halmos, B., 1956. Gondolatok Gyula város és környéke területrendezésével kapcsolatban. Földr. Ért. 5, 55-69.

Halmos, B., 1957. Békéscsaba és környéke településrendezésének kérdései. Földr. Ert. 6, 181-96.

Hartyányi, P.B. and Nováki, Gy., 1971. Angaben zum Beginn des neolitischen Pflanzenbaus im Südostungarischen Raum. Third Congress Intern. des Musées d'Agriculture Budapest 1971, 210-11.

Hartyányi, P.B. and Nováki, Gy., in press. Neolitikus gabonaleletek Dévaványáról. A Móra F. Múzeum Évkönyve.

Hartyányi, P.B., Nováki, Gy. and Patay, A., 1968. Növényi mag- és termelésleletek Magyarországon az ujkökortól a XVIII. századig. Magyar Mezögazd, Múz. Közleményei (1967-68), 5-82.

Helbaek, H., 1959. Domestication of food plants in the Old World. Science 130, 365-72.

Herrmann, M. et al., 1956. Ausgrabungen in der Petényi und Peskö Höhle (Bükk Gebirge). Fol. Arch. VIII, 3-22.

Higgs, E.S., 1962. Excavations at the early Neolithic site at Nea Nikomedeia, Greek Macedonia (1961 session) Part II. The biological data: Fauna. Proc. Prehist. Soc. 28, 271-74.

Higgs, E.S. and Vita-Finzi, C., 1972. Prehistoric economies: a territorial approach, in Higgs., E.S. (Ed.) Papers in Economic Prehistory, Cambridge, 27-36.

Hollendonner, F., 1933-34. A kopáncsi és kotacparti telepek faszeneinek anthrakotomiai vizsgálata. Dolgozatok Szeged IX-X, 303-05.

Hollendonner, F., 1935. Az Alföld östörténelem korabeli erdeinek meghatározása, anthrakotomiai vizsgálatok alapján. Math. Term. Tud. Ert. 59, 59-69.

Horváth, A., 1959. Az 1958. év régészeti kutatásai. Rég. Füz. 11, 16.

Horváth, A., 1972. Az 1971. év régészeti kutatásai. Rég. Füz. 25, 8.

Horváth, K., 1908. A Zalavöly település földrajza. Földr. Közl. XXXVI, 50-70.

Horváth, L., 1970. Az 1969. év régészeti kutatásai. Rég. Füz. 23, 8.

Hughes, R.E. 1958. Sheep population and environment in Snowdonia. Jour. Ecol. 46, 169-90.

Hutchinson, J., 1972. Farming and Food Supply. Cambridge.

Iversen, J., 1956. Forest clearance in the Stone Age. Scient. American 194 (3), 36-41.

Jankovich, M., 1967. A magyar szarvasmarha eredete és hasznositása. Agr. Szemle IX, 420-431.

Járai-Komlódi, M., 1966. Palinologiai vizsgálatok a Magyar Alföldön a Würm glacialis és a Holocén klima- és vegetációtörténetére vonatkozóan. Unpublished Ph.D. thesis, University of Budapest.

Járai-Komlódi, M., 1969. Adatok az Alföld negyedkori klima és vegetációtörténetéhez. II. Bot. Közl. 56, 43-55.

Járai-Komlódi, M., 1971. Palynological studies on swamps of the Zemplén Mountains. Ann. Univ. Scient. Budapest.. Sect. Biolog. 13, 103-13.

Jarman, H.N., 1972. The origins of wheat and barley cultivation, in Higgs, E.S. (Ed.) Papers in Economic Prehistory, Cambridge, 15-26.

Jenny, H., 1941. Factors of Soil Formation. New York.

Jósa, A., 1915. Ásatások a Gávai Katódombon és környéken. Arch. Ért. XXXV, 197-210.

Kadic, O., 1907. Adatok a szinvavölgyi diluviális ember kérdéséhez. Földt. Közl. 37, 333-45.

Kadic, O., 1914. Az 1913-ban végzett barlangkutatásaim eredményei. Barlangkutatás, 2, 189-90.

Kadic, O., 1916. A Büdöspestben 1916. évben végzett ásatás eredményei. Barlangkutatás 4, 136-40.

Kadic, O., 1933. A cserépfalui Mussolini barlang. Barlangkutatás 11.

Kadic, O., 1934. A jégkor embere Magyarországon. Földt. Int. Ért. 30, 60.

Kakas, J. (Ed.) 1960. Magyarország Éghajlati Atlasza. Budapest.

Kalicz, N., 1954. Kunszentmárton, co. Szolnok. Unpublished. In Record Section of Hungarian National Museum, no. 2117.

Kalicz, N., 1955. Adatok a harangalakú edények budapestkörnyéki elterjedéséhez. *Fol. Arch.* VII, 45-60.

Kalicz, N., 1956. Régészeti leletek és kutatások Szabolcs-Szatmár megye területén 1945-56 között. *Szabolcs-Szatmári Szemle,* 51-60.

Kalicz, N., 1957. Tiszazug Őskori települései. *Rég. Füz.* VIII.

Kalicz, N., 1957a. Leletmentő ásatások az 1955. évben. *Arch. Ért.* 84-85.

Kalicz, N., 1958. Az 1956. év régészeti kutatásai. *Rég. Füz.* 9, 14-15.

Kalicz, N., 1958a. Az 1957. év régészeti kutatásai. *Rég. Füz.* 10, 15.

Kalicz, N., 1958b. Régészeti kutatások és leletek Borsod-Abauj-Zemplén megyében 1958-b n. *A Herm. O. Múzeum Évkönyve* II, 24-32.

Kalicz, N., 1959. Az 1958. év régészeti kutatásai. *Rég. Füz.* 11, 27.

Kalicz, N., 1960. Az 1959. év régészeti kutatásai. *Rég. Füz.* 13, 59.

Kalicz, N., 1961. Az 1960. év régészeti kutatásai. *Rég. Füz.* 14, 26-27.

Kalicz, N., 1965. Siedlungsgeschichtliche Probleme der Körös- und Theiss Kultur. *Acta Antiqua et Arch.* VIII, 27-40.

Kalicz, N., 1966. Az 1965. év régészeti kutatásai. *Rég. Füz.* 19, 9.

Kalicz, N., 1967. Orosháza és környéke az őskorban, in Nagy, Gy. (Ed.) *Orosháza Földjének Története,* Orosháza, 93-102.

Kalicz, N., 1967a. Kőkori telep Aszódon. *Aszódi Múzeumi Füzetek,* 3, 33-47.

Kalicz, N., 1969. Heves megye régészeti emlékei I. *Magyarország Műemléki Topográfiája* VII, Budapest.

Kalicz, N., 1970. Tarnazsadány-Sándorrésze; Tiszavasvári-Keresztfal; Zalkod-Csenke, Jakabdomb; Hévizgyörk-Kaparóházi dűlő. *Mitt. des Arch. Inst. der Ungarischen Akad. der Wissenschaften* 1. Zehn Jahre Archäologische Forschung (1958-68), 134-35, 137, 139, 103-104.

Kalicz, N., and Makkay, J., 1966. Die Probleme der Linearkeramik im Alföld. *Acta Antiqua et Arch.* X, 35-47.

Kalicz, N., and Makkay, J., 1972. A medinai koraneolitikus leletek. *A Szekszárdi Balogh. A. Múzeum Füzetei* 10, 3-9.

Kalicz, N. and Makkay, J., 1972a. Probleme des Frühen Neolithikums der Nördliche Tiefebene, in *Die Aktuellen Fragen der Bandkeramik,* Akten der Pannonia Konferenzen I, Székesfehérvár, 77-81.

Kalicz, N. and Makkay, J., 1972b. Südliche Einflüsse im Frühen und Mittleren Neolithikum Transdanubiens, in *Die Aktuellen Fragen der Bandkeramik,* Akten der Pannonia Konferenzen I, Székesfehérvár, 93-97.

Kalicz, N., and Makkay, J., 1974. A méhteleki agyagistenek. *A Nyiregyházi Jósa A. Múzeum időszaki régészeti kiállitásának vezetője.*

Kalicz, N. and Makkay, J., 1974a. A méhteleki ujkőkori leletek. *Élet és Tudomány,* XXIX, 838-43

Kalicz, N. and Makkay, J., 1977. Die Linienbandkeramik in der Grossen Ungarischen Tiefebene. Budapest.

Kelemen, M., and Tőth, S., 1966. Az 1965. év régészeti kutatásai. Rég. Fűz. 19, 47.

Kemenczei, T., 1964. Az 1963. év régészeti kutatásai. Rég. Fűz. 17, 21.

Kemenczei, T., 1965. Az 1964. év régészeti kutatásai. Rég. Fűz. 18, 10.

Kemenczei, T., 1966. Az 1965. év régészeti kutatásai. Rég. Fűz. 19, 6.

Kemenczei, T., 1969. Az 1968. év régészeti kutatásai. Rég. Fűz. 22, 6-7.

Kemenczei, T. and Végh, K., 1964. A Herman Otto Múzeum leletmentései és ásatásai as 1959-63 évben. A Herm. O. Múzeum Évkönyve IV (1962-63), 233-42.

Kemenczei, T. and Végh, K., 1965. Régészeti munka a Borsod megyei szakaszon. Múzeumi Levelek Szolnok, 7-8, 24-25.

Kemenczei, T. and Végh, K., 1966. A Herman Otto Múzeum leletmentései és ásatásai az 1965. évben. A Herm. O. Múzeum Évkönyve VI, 403-07.

Kemenczei, T. and Végh, K., 1968. A Herman Otto Múzeum leletmentései és ásatásai az 1966. évben. A Herm O. Múzeum Évkönyve, VII, 391-99½

Kemenczei, T. and Végh, K., 1970. A Herman Otto Múzeum leletmentései és ásatásai az 1968. évben. A Herm.O. Múzeum Évkönyve IX, 411-14.

Kemenczei, T. and Végh, K., 1971. Leletmentések és ásatások az 1969. évben. A Herm. O. Múz. Évkönyve X, 507-15.

Király, I., 1963. A szarvasmarha tenyésztés átalakulása Somogy megyében 1848-1944. Agr. Szemle V, 177-207.

Kiss, A., 1956. Leletmentő ásatások az 1954. évben. Arch. Ért. 83, 96.

Kiss, A., 1958. Az 1957. év régészeti kutatásai. Rég. Fűz. 10, 8.

Kiss, A., 1961-62. Jelentés a balinkai avarkori temetőben végzett leletmentő ásatásakrol. Alba Regia II-III, 162-65.

Kiss, A., 1967. A tüskevári római telep. A Veszprém megyei Múzeumok Közl. 6, 37-52.

Kiss, L., 1929. A Földmüvelés a Rétközön. Debrecen.

Kiss, L. 1939. Fiatalabb kőkori telep és sirok Kenézlőn. Fol. Arch. I-II, 7-12.

Kisvarga, L., 1965. Sármellék gazdasági fejlődése 1731-1850 között. A Veszprém Megyei Múzeumok Közl. 3, 43-68.

Kocztur, E., 1964. Somogy megye régészeti leletkatasztere. Rég. Fűz. II, 13.

Kocztur, E., 1970. Az 1969. év régészeti kutatásai. Rég. Fűz. 23, 19.

Kőhegyi, M., 1961. Az 1960. év régészeti kutatásai. Rég. Fűz. 14, 18.

Kőhegyi, M., 1962. Az 1961. év régészeti kutatásai. Rég. Fűz. 15, 18.

Köppen, W. and Geiger, R., 1932. Handbuch der Klimatologie. Berlin.

Korek, J., 1951. Bükki kultúra telepe Oroson. Arch. Ért. 78, 68-72.

Korek, J. 1955. Das neolithische Fundmaterial der Höhle von Istál oskö. Acta Arch. Hung. 5, 141-43.

Korek, J., 1957. A vadnai neolitikus sirlelet. A Herm. O. Múzeum Évkönyve I, 14-30.

Korek, J., 1959. Az 1958. év régészeti kutatásai. Rég. Füz. 11, 22.

Korek, J., 1959a. Zu den anthropomorphen Darstellungen der Bükker Kultur. Fol. Arch. XI, 13-24.

Korek, J., 1960. A vonaldiszes kerámia kultúra elterjedése az Alföldön. A Móra F. Múzeum Évkönyve (1958-59), 19-52.

Korek, J., 1961. Neolithikus telep és sirok Dévaványán. Fol. Arch. XIII, 9-26.

Korek, J., 1962. Az 1961. év régészeti kutatásai. Rég. Füz. 15, 24.

Korek, J., 1963. Az 1962. év régészeti kutatásai. Rég. Füz. 16, 14.

Korek, J. 1964. Az 1963. év régészeti kutatásai. Rég. Füz. 17, 16, 20.

Korek, J., 1965. Ásatások Kiskörén. Múzeumi Levelek, Szolnok, 7-8, 5-15.

Korek, J., 1967. Az 1966. év régészeti kutatásai. Rég. Füz. 20, 16.

Korek, J., 1968. Die Linearkeramik auf dem Alföld. A Móra F. Múzeum Évkönyve, 13-20.

Korek, J., 1970. Nyiltszini bükki telep és sirok Aggteleken. Arch. Ért. 97, 3-22.

Korek, J., 1971. Az 1970. év régészeti kutatásai. Rég. Füz. 24, 15.

Korek, J., and Patay, P., 1956. A herpályi halom kökorvégi és rézkori települései. Fol. Arch. VIII, 23-42.

Korek, J. and Patay, P., 1958. A bükki kultura elterjedése Magyarországon. Rég. Füz. II, 2.

Köszegi, F., 1961. Az 1960. év régészeti kutatásai. Rég. Füz. 14, 22.

Kovalovszki, J., 1957. Régészeti adatok Szentes környéke település történetéhez. Rég. Füz. 5.

Kovalovszki, J., 1958. Az 1957 év. régészeti kutatásai, Rég. Füz. 10, 43-44.

Kovalovszki, J., 1964. Az 1963. év régészeti kutatásai. Rég. Füz. 17, 97.

Kozák, E., 1964. Atkár, Magyréde és Gyöngyöspata régészeti lelöhelyei. Az Egri Múz. Évkönyve II, 141-50.

Kralovánszky, A., 1958. Az 1956. év régészeti kutatásai. Rég. Füz. 9, 11.

Kralovánszky, A., 1961-62. A Balinka-Mecser pusztai avarkori sirok. Alba Regia, II-III, 166-172.

Kralovánsky, A. and Kralovánsky, J., 1960. Az 1959. év régészeti kutatásai. Rég. Füz. 13, 17.

Krecsmárik, E., 1911. Szarvas középkori és őskori leleteiről. Arch. Ért. XXX, 378-82.

Krecsmárik, E., 1915. A békésszarvasi őstelepek. Arch. Ért. XXXV, 11-43.

Krecsmárik, E., 1927. Ujabb kőkori kalaszrajzok a szarvasi őstelepről. Arch. Ért. XLI, 185-92.

Kruk, J., 1961. Próba rekonstrukcji naturalnych warunków rozwoju spoteczeństw neolitycznych na obszarze lessow Niecki Nidziańskiej. Sprawozdania Arch. XXIII, 259-84.

Kruk, J., 1972-73. Antropogeniczne przemiany krajobrazu wyzyn lessowych w neolicie. Acta Arch. Carpath. XII, 109-29.

Kruk, J., 1973. Stefy zasiedlenia i eksploata cji srodoviska na obszarze zachodniejczesci wyzyny mato Polskiej w Nedicie. Polish Arch. Abstracts, Vol. II.

Kulczycka-Leciejewiczowa, A., 1968. Ze studiów nad kultura ceramiki wstegowej w Polsce. Arch. Polski, XIII, 1-61.

Kutzián, I., 1944. A Körös Kultura. Budapest.

Kutzián, I., 1946. Ujabb neolitikus leletek a Magyar Történelmi Múzeumban. Magyar Muzeum, Budapest.

Kutzián, I., 1958. Pomáz Zdravlyak. Arch. Ért., 85, note 81.

Kutzián, I., 1958a. Az 1956. év régészeti kutatásai. Rég. Füz. 9, 7.

Kutzián, I., 1959. Az 1958. év régészeti kutatásai. Rég Füz. 11, 26-27.

Kutzián, I., 1963. The Copper Age Cemetery of Tiszapolgar-Basatanya. Budapest.

Kutzián, I., 1970. Sárazsadány-Akasztószer. Mitt. des Arch. Inst. der Ung. Akad. der Wissenschaften 1. Zehn Jahre Archaologische Forschung (1958-68), 129-30.

Lamb, H. H. et al., 1966. Atmospheric circulation and the main climatic variables between 8000 and 0 B.C.: meteorological evidence. World Climate from 8000 to 0 B.C. Royal Met. Soc. London, 174-217.

Lászlóffy, W., 1969. A Tisza, in A Tiszai Alföld. Magyarország tájföldrajza 2, Budapest, 11-26.

Lichardus, J., 1972. Zur Entestehung der Linearbandkeramik. Germania, 50, 1-15.

Lösch, A., 1954. The Economics of Location. New Haven.

Ložek, V., 1967. Climatic zones of Czechoslovakia during the Quaternary, in Cushing, E.J. and Wright, H.E. (Eds.) Quaternary Paleoecology, New Haven.

Makkay, J., 1957. A bihari Berettyóvölgy őskori leletei. A Debreceni Déri Múzeum Évkönyve (1948-56), 21-40.

Makkay, J., 1960. Az 1959. év régészeti kutatásai. Rég. Füz. 13, 15-16.

Makkay, J., 1968. Az 1967. év régészeti kutatásai. Rég. Füz. 21, 5.

Makkay, J., 1969. Az 1969. év régészeti kutatásai. Rég. Füz. 22, 19.

Makkay, J., 1969a. Zur Geschichte der Erforschung der Körös-Starčevo Kultur und einigen ihrer wichtigsten Probleme. Acta Arch. Hung. 21, 13-31.

Makkay, J., 1970. A kőkor és a rézkor Fejér megyében. Fejér megye története I/1. 14-36.

Makkay, J., and Trogmayer, O., 1966. Die bemalte Keramik der Körös Gruppe. A Móra F. Múz. Évkönyve (1964-65), 47-58.

Maksay, F., 1971. A Magyar Falu Középkori Településrendje. Budapest.

Maráz, B., 1972. Az 1971. év régészeti kutatásai. Rég. Füz. 25, 13.

Marosi, S., 1954. Geomorfológiai megfigyelések a Mezőföld Balatontól EK-re elterülő részén. Földr. Ért. 3, 433-43.

Márton, L., 1909. Egy sajátos kőkori edényalak. Arch. Ért. XXIX, 134-58.

Márton, L., 1912. Die wichtigsten Resultate vor- und frühgeschichtlichen Forschung in Ungarn. Prähist. Zeitsch. 3, 178-91.

Máté, F., 1955. Adatok tiszantúli réti talajaink genetikájához. Agr. és Talajt. 4, 133-43.

McAdams, R., 1965. Land Behind Baghdad—A History of Settlement in the Diyala Plains. Chicago.

Megay, G., 1960. Az 1959. év régészeti kutatásai. Rég. Füz. 13.

Megay, G., 1962. Az 1961. év régészeti kutatásai. Rég. Füz. 15, 22.

Méri, I., 1954. Beszámoló a tiszalök-rázompusztai és túrkeve-mórici ásatások eredményeiről. II. Arch. Ért. 81, 138-54.

Mesterházy, K., 1966. A Déri Múzeum régészeti tevékenysége 1962-65. A Debreceni Déri Múz. Évkönyve (1965), 19-55.

Mesterházy, K., 1966a. Beszámoló az 1965. évi Biharkeresztes-ártándi ásatásokról. A Debreceni Déri Múz. Évkönyve (1965), 61-67.

Mesterházy, K., 1970. A Déri Múzeum régészeti tevékenysége 1966-68. A Debreceni Déri Múz. Évkönyve (1968), 61-77.

Mihalek, J., 1897. A boldogkőváraljai neolitkori telepek. Arch. Közlemények 37.

Milojčić, V., 1949. Chronologie der jüngeren Steinzeit Mittel- und Südosteuropas. Berlin.

Milojčić, V., 1950. Körös-Starčevo-Vinča. Reinecke Festschrift. Mainz, 108-17.

Mithay, S., 1954. Leletmentő ásatások az 1952. évben. *Arch. Ért.* 81, 71.

Mithay, S., 1956. Leletmentő ásatások az 1954. évben. *Arch. Ért.* 83, 96.

Mithay, S., 1966. Zselizi tipusú leletek a Győr-Pápai Vám újabbkőkori lakótelepen. *Arrabona*, Győr 5-41.

Modderman, J.J.R., 1958-59. Die geographische Lage der bandkeramischen Siedlungen in den Niederlanden. *Paleohist.* VI-VII, 1-7.

Modderman, J.J.R., 1970. Linearbandkeramik aus Elsloo und Stein. *Analecta Praehist. Leidensia* III.

Móra, F., 1930. Ismeretlen tipusú edény Lebőről. *Dolgozatok Szeged* VI, 159-60.

Murray, J., 1970. *The First European Agriculture—A Study of the Osteological and Botanical Evidence until Two Thousand B.C.* Edinburgh.

Nagy, L., 1960. Az 1959. év régészeti kutatásai, *Rég. Füz.* 13, 2.

Nagy, M., 1954. Talajföldrajzi megfigyelések a Tiszazugban. *Földr. Ert.* 3, 507-543.

Nagy, M., 1956. Talajföldrajzi kutatások a két Körös melléken. *Földr. Közl.* 4, 1-20.

Nandris, J.G., 1970. Ground Water as a Factor in the First Temperate Neolithic Settlement of the Körös region. *Zbornik Belgrade Nat. Muz.* 6.

Nandris, J.G., 1972. Relations between the Mesolithic, the First Temperate Neolithic, and the Bandkeramik: the nature of the problem. In *Die Aktuellen Fragen der Bandkeramik,* Akten der Pannonia Konferenzen I, Székesfehérvár, 61-69.

Narr, K.J., 1956. Early food-producing populations, in Thomas, W.L. (Ed.) *Man's Role in Changing the Face of the Earth.* Chicago, 134-51.

National Atlas of Hungary, 1967. Budapest.

Németh, P., 1967. Szabolcs és Szatmár megyék Arpád kori (XI-XIII) századi földvárai és monostarai. *A Nyiregyházai Jósa A. Múz. Évkönyve,* X, 91-101.

Németh, P., 1972. Az 1971. év régészeti kutatásai. *Rég. Füz.* 25, 74.

Nepper, I., 1970. Megjegyzések a Körös csoport eszközkészletének vizsgálatához. *A Debreceni Déri Múz. Évkönyve* (1968), 79-109.

Nepper, I., 1972. Az 1971. év régészeti kutatásai. *Rég. Füz.* 25, 11.

Nepper, I. and Máthé, M., 1971. A Hajdu-Bihar megyei múzeumok régészeti tevékenysége 1969-71. *A Debreceni Déri Múz. Évkönyve,* 35-54.

Oates, D. and Oates, J., 1976. Early irrigation agriculture in Mesopotamia, in Sieveking, G de G., Longworth, I.M. and Wilson, K.E. (Eds.) *Problems in Economic and Social Archaeology*, London, 109-135.

Olasz, E., 1960. Az 1959. év régészeti kutatásai. Rég. Füz. 13, 28, 114-15.

Olasz, E., 1961. Az 1960. év régészeti kutatásai. Rég. Füz. 14, 16-17.

Olasz, E., 1962. Az 1961. év régészeti kutatásai. Rég. Füz. 15, 42, 89.

Olasz, E., 1963. Az 1962. év régészeti kutatásai. Rég. Füz. 16, 85.

Olasz, E., 1965. Az 1964. év régészeti kutatásai. Rég. Füz. 18, 40.

Olasz, E., 1970. Az 1969. év régészeti kutatásai. Rég. Füz. 23, 47.

Palov, J., 1957. Leletmentő ásatások az 1955 évben. Arch. Ért. 84, 84.

Papp, A., 1960. Fiatalkori vizrajzi változások a Tiszántul középső részében történelmi adatok alapján. A Debreceni K.L. Tud. Egy. Földr. Int. Közleményei, 42, 1-8.

Papp, L., 1966. Az 1965. év régészeti kutatásai. Rég. Füz. 19, 11.

Papp, L., 1967. Az 1966. év régészeti kutatásai. Rég. Füz. 20, 17.

Papp, L., 1970. Letkés-Keresztföldek. Mitt. des Arch. Inst. der Ungarischen Akad. der Wissenschaften 1. Zehn Jahre Archäologische Forschungen (1958-68), 110-11.

Párducz, M., 1949. Ásatási jelentések 1949. Bükk-hegység-Kölyuk II barlang. Arch. Ért. 76, 111-12.

Párducz, M., 1963. Az 1962. év régészeti kutatásai. Rég. Füz. 16, 22.

Párducz, M., 1964. Az 1963. év régészeti kutatásai. Rég. Füz. 17, 9.

Párducz, M., 1970. Felsőtárkány-Várhegy. Mitt. des Arch. Inst. der Ungarischen Akad. der Wissenschaften 1. Zehn Jahre Archäologische Forschung (1958-68), 100-101.

Patay, P., 1941. Kapcsolatok a Bükki és körösi kultura között. Arch. Ért. 1-3.

Patay, P., 1956. Ujabb kőkori és kelta leletek Nógradkövesden és a nógrádi dombvidéken. Arch. Ért. 83, 186-91.

Patay, P., 1957. A neolithikum a bodrogkeresztúri Kutyasoron. Fol. Arch. IX, 25-37.

Patay, P., 1961. A bodrogkeresztúri kultúra temetői. Rég. Füz. II, 10.

Patay, P., 1966-67. Adatok Budapest környékének ujkőkorákoz és rézkorához. Fol. Arch. XVIII, 7-26.

Patay, P., 1969. Az 1968. év régészeti kutatásai. Rég. Füz. 22, 18, 33.

Pavúk, J., 1973. Zur Chronologie und zu kulturellen Beziehungen der älteren Linearkeramik. Eight Congres ISPP Beograd 1971, 273-81.

Pécsi, M., 1959. A Magyarországi Dunavölgy Kialakulása és Felszinalaktana. Földrajzi Monográfiák III, Budapest.

Pécsi, M., 1967. A löszfeltárások üledékeinek genetikai osztályozása a Kárpát-medencében. Földr. Ért. 15, 1-17.

Pécsi, M. and Góczán, L., 1971. The soils of Hungary, in Pécsi, M., Enyedi, Gy., and Marosi, S. (Eds.), Hungary-Geographical Studies, Budapest, 125-34.

Pécsi, M. and Jakucs, P., 1971. The natural vegetation of Hungary, in Pécsi, M., Enyedi, Gy., and Marosi, S. (Eds.), Hungary-Geographical Studies, Budapest, 109-24.

Pécsi, M. and Sárfalvi, B., 1964. The Geography of Hungary. London.

Pécsi, M. and Somogyi, S., 1971. The hydrography of Hungary, in Pécsi, M., Enyedi, Gy., and Marosi, S. (Eds.), Hungary-Geographical Studies, Budapest, 85-107.

Pécsi, M., Somogyi, S. and Jakucs, P., 1971. Landscape units and their types in Hungary, in Pécsi, M., Enyedi, Gy., and Marosi, S. (Eds.), Hungary-Geographical Studies, Budapest, 11-64.

Péczely, Gy., 1967. Az Alföld éghajlatának jellemzői, in A Dunai Alföld. Magyarország tájföldrajza 1, Budapest, 34-47.

Pekáry, T., 1954. Leletmentő ásatások az 1954. évben. Arch. Ért. 81, 72.

Petres, E., 1954. Ujabb kőkori sírok Bicskén. Fol. Arch. VI, 22-28.

Petres, E., 1959. Neolithic graves at Bicske. István Király Múz. Közleményei 9, 5-14.

Petres, E., 1962. Az 1961. év régészeti kutatásai. Rég. Füz. 15, 4.

Petróczy, J., 1945. Korarézkori temető Tiszadobon. Fol. Arch. V, 35-43.

Piggott, S., 1965. Ancient Europe from the Beginnings of Agriculture to Classical Antiquity. Edinburgh.

Plog, F., 1971. Some operational considerations, in Gummerman, G.J. (Ed.) The Distribution of Prehistoric Population Aggregates. Prescott, 45-54.

Plog, F. and Hill, J.N., 1971. Explaining variability in the distribution of sites, in Gummerman, G. J. (Ed.) The Distribution of Prehistoric Population Aggregates. Prescott, 7-36.

Pusztai, R., 1958. Az 1957. év régészeti kutatásai. Rég. Füz. 10, 4.

Pusztai, R., 1962. Az 1961. év régészeti kutatásai. Rég. Füz. 15, 49.

Quitta, H., 1960. Zur Frage der ältesten Bandkeramik in Mitteleuropa. Praehist. Zeitschr. 38, 1-38, 153-88.

Quitta, H., 1964. Zur Herkunft des frühen Neolithikums in Mitteleuropa. Festschrift H. Unverzagt, Berlin, 14-24.

Quitta, H., 1969. Zur Deutung bandkeramischer Siedlungsfunde aus Auen und grundwassernahen Standorten. Deutsche Akad. der Wissenschaften zu Berlin, Band 25, 42-55.

Quitta, H., 1970. Zur Lage und Verbreitung der bandkeramischen Siedlungen im Leipziger Land. Zeitschr. für Archäologie 4, 155-76.

Reed, C. A., 1969. The pattern of animal domestication in the prehistoric Near East, in Ucko, P.J. and Dimbleby, G.W. (Eds.) The Domestication and Exploitation of Plants and Animals. London, 149-72.

Renfrew, C., 1972. The Emergence of Civilization—The Cyclades and the Aegean in the Third Millenium B.C. London.

Renfrew, J.M., 1969. The archaeological evidence for the domestication of plants: methods and problems, in Ucko, P.J. and Dimbleby, G.W. (Eds.) The Domestication and Exploitation of Plants and Animals. London, 149-72.

Renfrew, J.M., 1973. Agriculture, in Theocharis, D.R. (Ed.) Neolithic Greece.

Rónai, A., 1961. Az Alföld Telajviztérképe. Budapest.

Russell, R.J., 1967. River and Delta Morphology. Baton Rouge.

Saád, A., 1929. A Bükk hegységben végzett ujabb kutatások eredményei. Arch. Ért. 43, 238-47.

Saád, A., 1959. Ujabb gyüjtések eredményei a Korlat melletti Ravaszlyuktetön. Fol. Arch. XI, 3-12.

Saád, A. and Gaál, I., 1934. Elözetes jelentés a diósgyöri barlangban végzett ásatásokról. Barlangkutatás, 4, 12-19.

Sági, K., 1968. A Balaton vizállás tendenciái 1863-ig a történeti és kartográfiai adatok tükrében. A Veszprém Megyei Múzeumok Közleményei, 7, 441-462.

Salamon, A., 1960. Funde von NO-Ungarn aus der Römerzeit. Fol. Arch. XII, 145-59.

Salamon, A., 1962. Az 1961. év régészeti kutatásai. Rég. Füz. 15, 50.

Salamon, A., 1963. Az 1962. év régészeti kutatásai. Rég. Füz. 16, 46.

Salamon, A., 1966. Az 1965. év régészeti kutatásai. Rég. Füz. 18, 34.

Salamon, A., 1970. Szilvásvárad. Mitt. des Arch. Inst. der Ungarischen Akad. der Wissenschaften 1. Zehn Jahre Archäologische Forschung (1958-68), 142-43.

Sauer, C.O., 1962. Cultural geography, in Wagner, Ph.L. and Mikesell, M.W. (Eds.) Readings in Cultural Geography. Chicago.

Schupiter, E. (same as Zalotay, E.), 1931. Neolithikus telepek Szentes környéken. Dolgozatok Szeged VII, 54-59.

Schwarz, K., 1948. Lagen die Siedlungen der linearbandkeramischen Kultur Mitteldeutschlands in waldfreien oder in bewaldeten Landschaften? Strena Praehistorica, Martin John Festschr. Halle, 1-28.

Schwarzbach, M., 1961. Das Klima der Vorzeit. Stuttgart.

Selmeci, L., 1969. Das Wohnhaus der Körös Gruppe von Tiszajenö. Neuere Angaben zu den Haustypen des Frühneolithikums. A Móra F. Muzeum Évkönyve, 17-22.

Sherratt, A.G., 1972. Socio-economic and demographic models for the Neolithic and Bronze Ages of Europe, in Clarke, D.L. (Ed.) Models in Archaeology, London, 477-542.

Sherratt, A.G., 1973. The interpretation of change in European prehistory, in Renfrew, C., (Ed.) The Explanation of Culture Change: Models in Prehistory. London, 419-428.

Sielmann, B., 1971. Der Einfluss der Umwelt auf die neolithische Besiedlung Südwestdeutschlands unter besonderer Berücksichtigung der Verhältnisse am nördlichen Oberrhein. Acta Praehistorica et Archaeologica 2, 65-197.

Sielmann, B., 1972. Die frühneolitische Besiedlung Mitteleuropas. Fundamenta Reihe A, Band 3, Köln.

Smith, P.E.L.,1972. Land use, settlement patterns and subsistence agriculture: a demographic perspective, in Ucko, P.J., Tringham, R. and Dimbleby, G.W., Man, Settlement and Urbanism. London, 409-425.

Somogyi, S., 1962. A holocen időszakra vonatkozó kutatások földrajzi (hidromorfológiai) értékelése. Földr. Ért. 9, 185-202.

Somogyi, S., 1964. Geographical factors of the formation of alkali soils (Szik-soils) in Hungary, in Pécsi, M. (Ed.) Applied Geography, Budapest, 36-57.

Somogyi, S., 1965. A szikesek elterjedésének időbeli változásai Magyarországon. Földr. Közl. XIII, 41-55.

Somogyi, S., 1967. Ősföldrajzi és morfológiai kérdések az Alföldről. Földr. Ért. 16, 319-337.

Somogyi, S., 1971. Natural endowments of the Great Hungarian Plain, in Sárfalvi, B. (Ed.) The Changing Face of the Great Hungarian Plain, Budapest, 35-77.

Soó, R., 1931. A magyar puszta fejlődéstörténetének problémája. Földr. Közl. LIX, 1-15.

Soó, R., 1959. Az Alföld növényzete kialakulásának mai megítélése és vitás kérdései. Földr. Ért. 8, 1-18.

Sőregi, J., 1935. Jelentés a Déri Múzeum 1935. évi működéséről és állapotáról. A Debreceni Déri Múz. Évkönyve, 31, 67-69.

Soudsky, B. and Pavlů, I., 1972. The Linear Pottery culture settlement patterns of central Europe, in Ucko, P.J., Tringham, R. and Dimbleby, G.W. (Eds.) Man, Settlement and Urbanism, London 317-28.

Stanczik, I., 1969. Az 1968. év régészeti kutatásai. Rég. Füz. 22, 22.

Stanczik, I., 1970. Az 1969. év régészeti kutatásai. Rég. Füz. 23, 17.

Stefanovits, P., 1953. A szatmári sikság talajairól. Agr. és Talajtan 3, 19-33.

Stefanovits, P., 1969. Talajok, in A Tiszai Alföld. Magyarország tájföldrajza 2, Budapest, 61-66.

Stefanovits, P., 1971. Brown Forest Soils of Hungary. Budapest.

Steward, J.H., 1955. Theory of Culture Change–The Methodology of Multi-linear Evolution. Urbana.

Stieber, J., 1956. A Hillebrand Jenő barlang ujkőkori cölöpfa maradványai. Bot. Közl. XLVI, 287-90.

Sulimirski, T., 1961. The climate of the Ukraine during the Neolithic and Bronze Age. Archeologia XII, 1-17.

Szabó, J., 1962. Az 1961. év régészeti kutatásai. Rég. Füz. 15, 20, 22, 23.

Szabó, K., 1934. Kisréti part. Arch. Ért. XLVII, 9-30.

Szabolcs, I., 1961. A Vizrendezések és Öntözések Hatása a Tiszántuli Talajképződési Folyamatokra. Budapest.

Szabolcs, I., 1966. A Genetikus Üzemi Talajtérképezés Módszertankönyve.

Szeghalmy, Gy., 1913. Ásatás a szeghalmi Kovács-halomban. Arch. Ert., XXXIII, 123-41.

Székely, Gy., 1961. Vidéki termelőágak és as árúkereskedelem a XV-XVI században. Agr. Szemle III, 309-43.

Szendrei, J., 1883. Borsod megye őstelepei. Arch. Ért. III, 126-27.

Szentléleky, I., 1960. Az 1959. év régészeti kutatásai. Rég. Füz. 13, 14.

Szentmihályi, I., 1954. Leletmentő ásatások az 1952. évben. Arch. Ért. 81, 70.

Szücs, L., 1967. A Nagykunság talajai, különös tekintettel a csernozjomok képződésére. Agr. és Talajtan 16, 1-22.

Takács, L., 1967. A 'Rókafarku-köles' termelése a Nyugat Dunántulon. Agr. Szemle, IX, 297-307.

Tempir, Z., 1964. Beitrage zur ältesten Geschichte des Pflanzenbaus in Ungarn. Acta Arch. Hung. XVI, 65-98.

Thornbury, W.D., 1969. Principles of Geomorphology. New York.

Thran, P. and Broekhuizen, S., 1965. Agroecological atlas of cereal growing in Europe, in Agroclimatic Atlas of Europe. Pudoc, Centre for Agricult. Wageningen.

Tompa, F., 1929. Bandkeramik in Ungarn. Budapest.

Tompa, F., 1937. 25 Jahre Urgeschicht-Forschung in Ungarn 1912-36. Ber. Rom. Germ. Komm. XXIV-XXV (1934-35), 27-115.

Tompa, F., 1942. Budapest az Őskorban. Budapest.

Torma, I., 1968. Az 1967. év régészeti kutatásai. Rég. Füz. 21, 14.

Torma, I., 1969. Az 1968. év régészeti kutatásai. Rég. Füz. 22, 17.

Torma, I., 1970. Regöly-Füzfás, Tamásai. Mitt. des Arch. Inst. der Ung. Akad. der Wissenschaften 1. Zehn Jahre Arch. Forschung (1958-68), 127-29, 133.

Torma, I., 1970a. Az 1969. év régészeti kutatásai. Rég. Füz. 23, 16.

Torma, I., 1971. Neolithische Siedlung und Gräaberfeld, Kupferzeitliche Siedlung, Bronzezeitliche Siedlung in Pári-Altäcker. *Mitt. des Arch. Inst. der Ung. Akad. der Wissenschaften* 2, 27-34.

Torma, I., 1971a. Das Gräberfeld von Szakály aus der Mittleren Bronzezeit. *Mitt. des Arch. Inst. der Ung. Akad. der Wissenschaften* 2,

Tóth, E., 1967. Az 1966. év régészeti kutatásai. *Rég. Füz*. 20, 17.

Trbuhović, V., 1968. Pregled gustine naselenosti hroz praistoriju Srbije. *Starinar* XIX, 46.

Trewartha, G.T., 1954. *An Introduction to Climate*. New York.

Tringham, R., 1968. A preliminary study of the early neolithic and latest mesolithic blade industries in Southeast and Central Europe, in Coles, J.M. and Simpson, D.D.A. (Eds.) *Studies in Ancient Europe*, Leicester, 45-70.

Tringham, R., 1971. *Hunters, Fishers and Farmers of Eastern Europe 6000-3000 B.C.* London.

Trogmayer, O., 1958. Az 1956. év régészeti kutatásai. *Rég. Füz*. 9, 13-14.

Trogmayer, O., 1960. Az 1959. év régészeti kutatásai. *Rég. Füz*. 13, 14.

Trogmayer, O., 1965. Megjegyzések a Körös csoport relativ idörendjéhez. *Arch. Ért*. 91, 67-86.

Trogmayer, O., 1965. Az 1964. év régészeti kutatásai. *Rég. Füz*. 18, 9, 18.

Trogmayer, O., 1966. Az 1965. év régészeti kutatásai. *Rég. Füz*. 19.

Trogmayer, O., 1967. Az 1966. év régészeti kutatásai. *Rég. Füz*. 20, 10.

Trogmayer, O., 1968. A Dél-Alföld korai Neolitikumának föbb kérdései. Unpublished Ph.D. thesis, Szeged.

Trogmayer, O., 1968a. A Körös csoport barbotin kerámiájáról. *Arch. Ért.* 95, 6-12.

Trogmayer, O., 1971. Tápé területének régészeti emlékei. In *Tápé Története és Néprajza*, 31-45.

Uzsoki, A. and Gabler, D., 1967. Az 1966. év régészeti kutatásai. *Rég. Füz*. 20, 20.

Vadász, E., 1969. Zur praehistorischen Siedlungs- und Klimageschichte des Bezirks von Kalocsa. *A Móra F. Múz. Évkönyve*, 83-92.

Vadász, E., 1972. Az 1971. év régészeti kutatásai. *Rég. Füz.* 25, 3.

Valter, I., 1969. Az 1968. év régészeti kutatásai. *Rég. Füz.* 22, 68.

Van Zeist, W., 1969. Reflections on prehistoric environments in the Near East, in Ucko, P.J., and Dimbleby, G.W. (Eds.) *The Domestication and Exploitation of Plants and Animals*. London, 35-46.

Végh, K., 1970. Az 1969. év régészeti kutatásai. *Rég. Füz*. 23, 3.

Végh, K. and Kemenczei, T., 1965. A Hermann Ottó Múzeum leletmentései és ásatásai 1964-ban. *A Hermann O. Múz. Évkönyve* (1964-65), 449-53.

Vértes, L., 1953. Az alsópaleolitikum embereinek első biztos eszközkészlete Magyarországon. *Arch. Ért.* 80, 17-26.

Vértes, L., 1965. Zur Technologie Grobgeratiger Silexfunde in Nordungarn. *Fol. Arch.* XVII, 9-36.

Visegrádi, J., 1907. Festett cesrépedény töredékek a Sátoraljaújhely őstelepről. *Arch. Ért.* XXVII, 279-87.

Visegrádi, J., 1912. A Sátoraljaújhelyi őstelep. *Arch. Ért.* XXXII, 244-61.

Vita-Finzi, C. and Higgs, E.S., 1970. Prehistoric economy in the Mount Carmel area of Palestine. Site Catchment Analysis. *Proc. Prehist. Soc.* 36, 1-37.

Vozáry, E., 1957. Pollenanalytische Untersuchung des Torfmoores 'Nyirestó' im Nordosten der Ungarischen Tiefebene (Alföld). *Acta Botanica Akad. Scient. Hung.* III, 123-34.

Wallner, E., 1926. A felsőörvidéki magyarság települése. *Földr. Közl.* LIV, 1-31.

Walter, H., 1973. *Vegetation of the Earth–In relation to Climate and the Eco-Physiological Conditions.* London.

Walterbolk, H.T., 1971. Food production in prehistoric Europe, in Struever, S. (Ed.) *Prehistoric Agriculture,* New York, 335-58.

Wilhelmy, H., 1950. Das Alter der Schwarzerde und der Steppen Mittel- und Osteuropas. *Erdkunde* 4, 5-34.

Wiślański, T. (Ed.) 1970. *The Neolithic of Poland.*

Wosinsky, M., 1896. *Tolna Vármegye az Őskortól a Honfoglalásig.* Budapest.

Zalotay, E., 1932. Csongrád vármegye őskori települése. *Dolgozatok Szeged* VIII, 49-102.

Zohåry, D., 1969. The progenitors of wheat and barley in relation to domestication and agricultural dispersal in the Old World, in Ucko, P.J. and Dimbleby, G.W. (Eds.) *The Domestication and Exploitation of Plants and Animals,* London, 47-66.

Zoltai, L., 1907. Jelentés a Debreceni sz. kir. város Múzeuma 1906. évi működéséről és állapotárol. *A Debreceni Múzeum Kiadványa* 24-39.

Zoltai, L., 1910. Jelentés Debrecen sz. kir. város Múzeum 1909. évi működéséről és állapotáról. *A Debreceni Múzeum Kiadványai,* 15.

Zoltai, L., 1911. Jelentés Debrecen sz. kir. város Múzeuma 1910. évi működéséről állapotáról. *A Debreceni Múzeum Kiadványai,* 24-25; 36-48.

Zoltai, L., 1927. A Toczóvölgyi László és Szántay halmok megásatása. *A Debreceni Múzeum Kiadványa,* 47-52.

Zólyomi, B., 1953. Die Entwicklungsgeschichtliche der Vegetation Ungarns seit dem letzten Interglazial, _Acta Biol. Hung._ 4, 367-430.

Zólyomi, B., 1958. A növénytakaró történeti kialakulása, in Pécsi, M. (Ed.) _Budapest Természeti Képe_, Budapest.

Zólyomi, B., 1964. Pannonische Vegetationprobleme. _Verhand. der Zool.-Bot. Gesellschaft_ in Wien, 103, 144-51.

Zólyomi, B., 1971. 6000-jährige Geschichte der Agrikultur in der Umgebung des Balaton-Sees aufgrund von pollenstatistischen Untersuchungen der Seesedimenten. _Third Congres Int. des Musees d'Agriculture Budapest_, 194-95.

Zólyomi, B. and Simon, T., 1969. Természetes növényzet, in _A Tiszai Alföld_. Magyarország tájföldrajza 2, Budapest, 124-31.

APPENDIX A: SITE LIST

* settlements included in the analysis

Körös culture

*1. Alpár-Tiszapart, co. Bács-Kiskun. (Kutzián, 1944; Nepper, 1970)

2. Baks-Szobori út, co. Csongrád (Trogmayer, 1968)

3. Békés-Irányi Dániel u. iskola kertje, co. Békés (Nepper, 1970)

4. Békés-Körös part, co. Békés (Nepper, 1970)

5. Békésszentandrás, co. Békés (Csalog, 1958; Nepper, 1970)

*6. Békésszentandrás-Körösön túli része, co. Békés (Kalicz and Makkay, 1977, same as Szarvas I)

*7. Berettyóújfalu-Levente u., co. Hajdu-Bihar (Kutzián, 1944; Makkay, 1957; Nepper, 1970)

*8. Bokros-Buzáspart, co. Csongrád (Kalicz, 1960)

*9. Csánytelek-Tápai János földje, co. Bács-Kiskun (Kutzián, 1944; Kovalovszki, 1957)

*10. Csátalja, co. Bács-Kiskun (Kutzián, 1944; Nepper, 1970)

*11. Csépa-Csipsárpart III-Szira tanya, co. Szolnok (Kalicz, 1957)

*12. Csépa-Csipsárpart V-Dancza tanya, co. Szolnok (Kalicz, 1957; Nepper, 1970)

*13. Csongrád, co. Csongrád (Kovalovszki, 1957; Nepper, 1970)

*14. Csongrád-Belváros, co. Csongrád (Banner, 1932; Zalotay, 1932; Kutzián, 1944; Kovalovszki, 1957; Nepper, 1970)

*15. Derekegyháza, co. Csongrád (Kovalovszki, 1957; Nepper, 1970)

*16. Deszk C, co. Csongrád (Foltiny, 1941; Banner, 1932; Kutzián, 1944; Trogmayer, 1968)

17. Deszk G, co. Csongrád
(Nepper, 1970)

18. Deszk I, co. Csongrád
(Nepper, 1970)

*19. Deszk-Okapi dülö, co. Csongrád
(Nepper, 1970)

*20. Deszk-1 sz. olajkút, co. Csongrád
(Trogmayer, 1967; 1968)

*21. Dévaványa-Atyaszeg-Hajashalom, co. Békés
(Trogmayer, 1968)

*22. Dévaványa-Katalszeg, co. Békés
(Trogmayer, 1968)

*23. Dévaványa-Katonaföldek, co. Békés
(Escedy, 1971; 1972)

*24. Dévaványa-Pusztaecseg, co. Békés
(Kutzián, 1944; Trogmayer, 1968)

*25. Dévaványa-Réheli gát, co. Békés
(Gazdapusztai, 1959; Nepper, 1970; Ecsedy, 1971; 1972)

*26. Dévaványa-Simasziget, co. Békés
(Korek, 1963; Nepper, 1970)

*27. Dévaványa-Szikhalom, co. Békés
(Kovalovszki, 1957; Nepper, 1970)

*28. Dóc-Sziget déli része, co. Bács-Kiskun
(Trogmayer, pers. comm.)

29. Dusnok, co. Bács-Kiskun
(Trogmayer, 1968)

*30. Endröd, co. Békés
(Banner, 1932; Kutzián, 1944; Nepper, 1970)

*31. Felgyö-Ujmajor, co. Csongrád
(Bálint and Szabó, 1955; Nepper, 1970)

*32. Furta-Csátó-Vereshalom, co. Hajdu-Bihar
(Nepper, 1970)

*33. Gyálarét-Szilágyi major, co. Csongrád
(Trogmayer, 1965; 1968)

*34. Gyoma, co. Békés
(Kutzián, 1944)

*35. Gyoma-Csudaballa, co. Békés
(Kutzián, 1944; Nepper, 1970)

36. Gyoma-Nagyszirt, co. Békés
(Kutzián, 1944; Nepper, 1970)

*37. Gyula-Vasuti homokbánya, co. Békés
(Banner, 1932; Kutzián, 1944; Nepper, 1970)

*38. Hajós-Csipsa puszta, co. Bács-Kiskun
(Trogmayer, 1968)

39. Hajós-Hildi beköttöût, co. Bács-Kiskun
(Trogmayer, 1968)

*40. Hódmezövásárhely-Bodzáspart- Banga Imre tanya, co. Csongrád
(Banner, 1940; Nepper, 1970)

*41. Hódmezövásárhely-Bodzáspart-Pap bognár földje, co. Csongrád
(Banner, 1937; 1938; Kutzián, 1944; Nepper, 1970)

*42. Hódmezövásárhely-Fecskés Muzsik tanya, co. Csongrád
(Kutzián, 1944; Nepper, 1970)

*43. Hódmezövásárhely-Gorzsa-Kovács István tanya, co. Csongrád
(Gazdapusztai, 1957; Nepper, 1970)

*44. Hódmezövásárhely-Gorzsa-Pócsy tanya, co. Csongrád
(Banner, 1937; Kutzián, 1944; Nepper, 1970)

*45. Hódmezövásárhely-Gorzsa-Vermes tanya, co. Csongrád
(Banner, 1932; Kutzián, 1944; Nepper, 1970).

*46. Hódmezövásárhely-Hámszaritódomb, co. Csongrád
(Zalotay, 1932; Banner, 1937; Kutzián, 1944; Nepper, 1970)

*47. Hódmezövásárhely-Kopáncs-Kovács tanya, co. Csongrád
(Banner, 1932; Kutzián, 1944; Nepper, 1970)

48. Hódemzövásárhely-Kopáncs-Lenin TSZ (Szenti tanya), co. Csongrád
(Gazdapusztai, 1966; Nepper, 1970)

*49. Hódmezövásárhely-Kopáncs-Nagy Szabó Pál tanya, co. Csongrád
(Banner, 1937; Kutzián, 1944; Nepper, 1970)

*50. Hódmezövásárhely-Kopáncs-Técsy tanya, co. Csongrád
(Banner, 1937; Kutzián, 1944; Nepper, 1970)

*51. Hódmezövásárhely-Kopáncs-Zsoldos tanya, co. Csongrád
(Banner, 1932; Kutzián, 1944; Nepper, 1970)

*52. Hódmezövásárhely-Kopáncsi rét, co. Csongrád
(Banner, 1940; Kutzián, 1944; Nepper, 1970)

*53. Hódmezövásárhely-Kotacpart-Vata tanya, co. Csongrád
(Banner, 1932; Kutzián, 1944; Nepper, 1970)

*54. Hódmezövásárhely-Nagysziget-Gál Imre tanya, co. Csongrád
(Banner, 1938; Kutzián, 1944; Nepper, 1970)

*55. Hódmezövásárhely-Nagysziget-Ujhelyi tanya, co. Csongrád
(Zalotay, 1932; Banner, 1940; Kutzián, 1944; Nepper, 1970)

*56. Hódmezövásárhely-Solt pálé-Égetö Bálint földje, co. Csongrád
(Banner, 1937; Kutzián, 1944; Nepper, 1970)

57. Kardoskút-480. tanya, co. Békés
(Kalicz, 1967; Nepper, 1970)

58. Kardoskút-Aranyadhalmi dülö, co. Békés
(Olasz, 1962; Kalicz, 1967; Nepper, 1970)

59. Kardoskút-Sportpálya, co. Békés
(Kalicz, 1967; Nepper, 1970)

*60. Kardoskút-Szeghalmi tanya, co. Békés
(Kalicz, 1967; Nepper, 1970)

*61. Kenderes-Kulis, co. Szolnok
(Trogmayer, 1968)

*62. Kengyelsziget-Délibáb u., co. Szolnok
(Trogmayer, 1968)

*63. Kúnszentmárton-Érpart I, co. Szolnok
(Kalicz, 1954)

*64. Kunszentmárton-Érpart V, co. Szolnok
(Kalicz, 1954)

*65. Kunszentmárton-Érpart VII, co. Szolnok
(Kalicz, 1954)

*66. Kunszentmárton-Érpart-Bábocka II, co. Szolnok
(Kalicz, 1954)

*67. Kunszentmárton-Érpart-Bábocka IV, co. Szolnok
(Kalicz, 1954)

*68. Kunszentmárton-Hidfö-Kékes VII, co. Szolnok
(Kalicz, 1957)

*69. Kunszentmárton-Kisjaksorér I, co. Szolnok
(Schupiter, 1931; Banner, 1932; Zalotay, 1932; Kutzián, 1944;
Kalicz, 1954; Nepper, 1970)

*70. Kunszentmárton-Kisjaksorér II, co. Szolnok
(Kalicz, 1954)

*71. Kunszentmárton-Kökényzug I, co. Szolnok
(Kalicz, 1954)

*72. Kunszentmárton-Kungyalu-Körtvélyes I, co. Szolnok
(Kalicz, 1957; Nepper, 1970)

*73. Kunszentmárton-Nagyérpart, co. Szolnok
(Kalicz, 1954)

*74. Kunszentmárton-Nagyjaksorér I, co. Szolnok
(Schupiter, 1931; Banner, 1932; Kalicz, 1954; Nepper, 1970)

*75. Kunszentmárton- Nagyjaksorér II, co. Szolnok
(Kalicz, 1954)

*76. Kunszentmárton-Péterszög, co. Szolnok
(Kalicz, 1954)

*77. Magyartés, co. Csongrád
(Banner, 1932; Kutzián, 1944; Nepper, 1970)

*78. Maroslele-Pana, co. Csongrád
(Trogmayer, 1964; 1968; Nepper, 1970)

*79. Mártély-Korhány halom, co. Csongrád
(Kutzián, 1944; Nepper, 1970)

*80. Mártély-Maczelka tanya, co. Csongrád
(Banner, 1937; Kutzián, 1944; Nepper, 1970)

81. Mártély-TSZ építkezés, co. Csongrád
(Trogmayer, 1968)

82. Mezőberény-Tücsökhalom, co. Békés
(Maráz, 1972)

*83. Mezőtúr-Berettyógát, Vasuti hid, co. Szolnok
(Banner, 1942; Trogmayer, 1968)

*84. Mezőtúr-Kovácshalom, co. Szolnok
(Banner, 1931; Kutzián, 1944; Nepper, 1970; Kalicz and Makkay, 1977, same as site 6 (Szarvas 1))

85. Mindszent, co. Csongrád
(Kutzián, 1944; Nepper, 1970)

*86. Nagyrév-Zsidóhalom IV, co. Szolnok
(Kalicz, 1957; Nepper, 1970)

*87. Nagyrév-Zsidóhalom V, co. Szolnok
(Kalicz, 1957; Nepper, 1970)

*88. Öcsöd-Daru, co. Szolnok
(Trogmayer, 1968)

*89. Öcsöd-Hidfő, co. Szolnok
(Kalicz, 1957; Nepper, 1970)

*90. Öcsöd-Kenderes halom, co. Szolnok
(Csalog, 1959; Nepper, 1970)

*91. Öcsöd-Kéthalom, co. Szolnok
(Korek, 1959)

*92. Öcsödi tanyák, co. Szolnok
(Kalicz, 1954)

*93. Orosháza-Bonum, co. Békés
(Olasz, 1962)

*94. Orosháza-Cinkus, co. Békés
(Olasz, 1965)

95. Pusztaföldvár, co. Békés
(Banner, 1932; Kutzián, 1944; Nepper, 1970)

*96. Pusztaföldvár-Bagi féle malom, co. Békés
(Olasz, 1962; Nepper, 1970)

97. Pusztaföldvár-Kunos Ferenc tanya, co. Békés
(Olasz, 1960; 1961; Kalicz, 1967; Nepper, 1970)

98. Pusztaföldvár-Vörösmarty u. 83., co Békés
(Olasz, 1961)

*99. Pusztaistvánháza, co. Szolnok
(Kalicz, 1957; Nepper, 1970)

100. Rákóczifalva, co. Szolnok
(Trogmayer, 1968)

*101. Röszke-Lüdvár, co. Csongrád
(Trogmayer, 1965; 1966; 1968)

*102. Röszke-Vasutállomás, co. Csongrád
(Nepper, 1970)

*103. Sövényháza, co. Csongrád
(Kutzián, 1944; Nepper, 1970)

*104. Sövényháza-Bibicshát, co. Csongrád
(Zalotay, 1932; Kutzián, 1944; Nepper, 1970)

*105. Sövényháza-Csontospart, co. Csongrád
(Trogmayer, pers. comm.)

*106. Sövényháza-Kötörés, co. Csongrád
(Banner, 1932; Zalotay, 1932; Kutzián, 1944; Nepper, 1970)

107. Szakmár, co. Bács-Kiskun
(Trogmayer, 1968)

*108. Szakmár-Kisülés, co. Bács-Kiskun
(Trogmayer, 1968)

*109. Szarvas-Halásztelek, co. Békés
(Kutzián, 1944; Nepper, 1970)

*110. Szarvas-Halez tanya, co. Békés
(Banner, 1931; 1932; Kutzián, 1944; Nepper, 1970)

111. Szarvas-Petöfi S. ut. 31, co. Békés
(Palov, 1957; Nepper, 1970)

*112. Szarvas-Szappanos, co. Békés
(Krecsmárik, 1915; 1927; Banner, 1931; 1932; Kutzián, 1944;
Nepper, 1970)

*113. Szegvár-Belterület, co. Csongrád
(Kutzián, 1944; Nepper, 1970)

114. Székkutas-Kishalom-Hármashatár, co. Csongrád
(Trogmayer, 1968)

*115. Szelevény-Brena, co. Szolnok
(Kalicz, 1957; Nepper, 1970)

*116. Szelevény-Hosszuhát, co. Szolnok
(Kalicz, 1957; Nepper, 1970)

*117. Szelevény-Kisasszonypart I, co. Szolnok
(Kalicz, 1957; Nepper, 1970)

*118. Szelevény-Kisasszonypart II, co. Szolnok
(Kalicz, 1957; Nepper, 1970)

119. Szelevény-Othalom, co. Szolnok
(Kalicz, 1957; Nepper, 1970)

*120. Szelevény-Pecekpart, co. Szolnok
(Kutzián, 1944; Kalicz, 1957; Nepper, 1970)

*121. Szelevény-Vadas-Hottinger tanya, co. Szolnok
(Kalicz, 1957; Nepper, 1970)

122. Szentes-Berek, co. Csongrád
(Banner, 1932; Kutzián, 1944; Kovalovszki, 1957; Nepper, 1970)

123. Szentes-Boldizsár tanya, co. Csongrád
(Csalog, 1966; Nepper, 1970)

*124. Szentes-Csucsföld, co. Csongrád
(Schupiter, 1931; Banner, 1932; Kovalovszki, 1957; Nepper, 1970)

*125. Szentes-Ficsorhalom, co. Csongrád
(Schupiter, 1931; Banner, 1932; Kovalovszki, 1957; Nepper, 1970)

*126. Szentes-Ilonapart, co. Csongrád
(Kovalovszki, 1957; Csalog, 1961; 1964; Nepper, 1970)

*127. Szentes-Jámborhalom, co. Csongrád
(Banner, 1932; Nepper, 1970)

*128. Szentes-Kistöke, co. Csongrád
(Schupiter, 1931; Zalotay, 1932; Kutzián, 1944; Nepper, 1970)

*129. Szentes-Kistöke-Dinnyéshalom, co. Csongrád
(Nepper, 1970)

130. Szentes-Kiszugi dülö, co. Csongrád
(Kutzián, 1944; Nepper, 1970)

*131. Szentes-Kurcapart, co. Csongrád
(Kutzián, 1944; Csalog, 1960; Trogmayer, 1968)

*132. Szentes-Nagyhegy-Gyarmati föld, co. Csongrád
(Kovalovszki, 1957; Trogmayer, 1968)

*133. Szentes-Nagyhegy-Muza János földje, co. Csongrád
(Schupiter, 1931; Kutzián, 1944; Kovalovszki, 1957)

134. Szentes-Nagyhegy-Sinóros Szabó földje, co. Csongrád
(Kutzián, 1944; Trogmayer, 1968)

135. Szentes-Nagytöke, co. Csongrád
(Kutzián, 1944)

*136. Szentes-Sáphalom, co. Csongrád
(Kovalovszki, 1957; Nepper, 1970)

137. Szentes-Szentlászló, co. Csongrád
(Kutzián, 1944; Trogmayer, 1968)

*138. Szentes-Tőke-Vekerpart, co. Csongrád
(Kovalovszki, 1957; Nepper, 1970)

*139. Szentes-Tüzköves, co. Csongrád
(Schupiter, 1931; Banner, 1932; Kutzián, 1944; Nepper, 1970)

140. Szentes-Vekerzug, co. Csongrád
(Kutzián, 1944)

*141. Szeremle-Görönd, co. Bács-Kiskun
(Kőhegyi, 1961; 1962; Nepper, 1970)

142. Szolnok-Tiszapart, co. Szolnok
(Kutzián, 1944)

143. Szőreg-Téglagyár, co. Csongrád
(Trogmayer, 1968)

*144. Tápé-Sirhegy, co. Csongrád
(Trogmayer, 1968)

145. Tápé-Vetyehát, co. Csongrád
(Trogmayer, 1971)

146. Telekpuszta, co. Békés
(Nepper, 1970)

147. Tenyői gát, co. Szolnok
(Trogmayer, 1968)

148. Tiszaföldvár, co. Szolnok
(Kalicz, 1957; Nepper, 1970)

*149. Tiszagyenda-Garahalom, co. Szolnok
(Trogmayer, 1968)

*150. Tiszainoka-Szigetpart II, co. Szolnok
(Kalicz, 1957; Nepper, 1970)

*151. Tiszainoka-Szőlősziget partja II, co. Szolnok
(Kalicz, 1957; Nepper, 1970)

*152. Tiszainoka-Szőlőút, co. Szolnok
(Kalicz, 1957; Nepper, 1970)

*153. Tiszajenő-Szárazérpart, co. Szolnok
(Selmeci, 1969)

*154. Tiszakürt-Műút melléke II, co. Szolnok
(Kalicz, 1957; Nepper, 1970)

*155. Tiszakürt-Tópart-Földesér, co. Szolnok
(Kalicz, 1957; Nepper, 1970)

*156. Tiszasas-Kőrösi Imre kertje, co. Szolnok
(Kalicz, 1957; Nepper, 1970)

*157. Tiszasas-Rév, co. Szolnok
(Kalicz, 1957; Nepper, 1970)

*158. Tiszasziget (Ószentiván) VIII, co. Csongrád
(Banner, 1932; Kutzián, 1944; Nepper, 1970)

*159. Tiszaszőlős-Aszó tanya, co. Szolnok
(Trogmayer, 1968)

*160. Tiszaug-Keménytető II, co. Szolnok
(Kutzián, 1944; Kalicz, 1957; Nepper, 1970)

*161. Tiszaug-Tópart III, co. Szolnok
(Banner, 1932; Kutzián, 1944; Kalicz, 1957; Nepper, 1970)

*162. Tószeg-Kucorgó, co. Szolnok
(Kutzián, 1944; Nepper, 1970)

*163. Tótkomlós-Gyulamező, co. Békés
(Olasz, 1963; 1970)

164. Tótkomlós-Viharsarok TSZ földje, co. Békés
(Olasz, 1960; Kalicz, 1967; Nepper, 1970)

*165. Vaskút-Hiesl kert, co. Bács-Kiskun
(Kutzián, 1944; Nepper, 1970)

*166. Vésztő-Kőtpuszta All. Gazd., co. Békés
(Kovalovszki, 1958; Nepper, 1970)

*167. Zalota, co. Csongrád
(Schupiter, 1931; Kutzián, 1944; Nepper, 1970)

*168. Zsáka-Várdomb, co. Hajdu-Bihar
(Makkay, 1957;, Kalicz, 1958; Nepper, 1970)

*169. Zsáka-Vekerdi ut, co. Hajdu-Bihar
(Trogmayer, 1968)

Alföld Linear Pottery culture

No.	Site	Chronology
* 1.	Abádszalók-Berei-Rév, co. Szolnok (Csalog, 1965; Korek, 1968; 1970; Kalicz and Makkay, 1977)	Early
2.	Abaújdevecser, co. Borsod (Korek, 1957; Korek and Patay, 1958; Kalicz and Makkay, 1977)	
* 3.	Abaújkér-Aranyos patak partja, co. Borsod (Korek and Patay, 1958; Korek, 1960; Kalicz and Makkay, 1977)	Late AVK, Tiszadob
4.	Abaújszántó, co. Borsod (Korek and Patay, 1958)	
* 5.	Abaújszántó-Mezőgazd. Techn. Diákotthona, co. B. (Kalicz and Makkay, 1977)	Late AVK, Tiszadob
6.	Abod, co. Borsod (Korek and Patay, 1958; Kalicz and Makkay, 1977)	
7.	Abony, co. Pest ((Kalicz and Makkay, 1977)	
* 8.	Aggtelek-Baradla barlang, co. Borsod (Márton, 1912; Tompa, 1929; Csalog, 1941; Korek and Patay, 1958; Kalicz and Makkay, 1977)	Late AVK, Tiszadob, Bükk II
* 9.	Aggtelek-Bradla barlang előtere, co. Borsod (Korek, 1970; Kalicz and Makkay, 1977)	Late AVK, Tiszadob, Bükk I
*10.	Alsózsolca-Házgyár, co. Borsod (Kemenczei and Végh, 1968; Kalicz and Makkay, 1977)	Early
*11.	Apagy-Peckesrét, co. Szabolcs (Kalicz and Makkay, 1977)	Tiszadob
*12.	Arnót-Homokbánya, co. Borsod (Végh, 1970; Kemenczei and Végh, 1971; Kalicz and Makkay, 1977)	Tiszadob Bükk I
*13.	Artánd-Lencsésdomb, co. Hajdu-Bihar (Kalicz and Makkay, 1977)	Middle
14.	Baktakék, co. Borsod (Korek and Patay, 1958)	
15.	Balmazújváros-Daru, co. Hajdu-Bihar (Kalicz and Makkay, 1977)	

*16.	Balmazújváros-Hortobágy-Árkuspart, co. H-B (Zoltai, 1910; Korek, 1960)	Szilmeg
*17.	Balmazújváros-Hortobágy-Csécsi Halastó, co. H-B (Mesterházy, 1966a; Kalicz and Makkay, 1977)	Middle
*18.	Balmazújváros-Hortobágy-Faluvéghalom, co. H-B (Gazdapusztai, 1964; Mesterházy, 1966a; Korek, 1970; Kalicz and Makkay, 1977)	Early
*19.	Balmazújváros-Hortobágy-Féllaponyag, co. H-B (Korek, 1960; 1961; Kalicz and Makkay, 1977)	Middle
*20.	Balmazújváros-Hortobágy-Gyökérkút, co. H-B (Korek, 1960; Kalicz and Makkay, 1977)	
*21.	Balmazújváros-Hortobágy-Kandrahalom, co. H-B (Zoltai, 1910; 1911; Korek, 1960; Kalicz and Makkay, 1977)	Middle
*22.	Balmazújváros-Hortobágy-Papegyháza, co. H-B (Zoltai, 1911; Korek, 1960; Kalicz and Makkay, 1977)	Middle
*23.	Balmazújváros-Hortobágy-Pipáshalom, co. H-B (Zoltai, 1911; Korek, 1960; 1968; Kalicz and Makkay, 1977)	Middle
*24.	Balmazújváros-Hortobágy-Szásztelek, co. H-B (Nepper, 1971)	Middle
*25.	Balmazújváros-Hortobágy-Viztároló, co. H-B (Sőregi, 1935; Korek, 1960; Kalicz and Makkay, 1977)	Middle
*26.	Balmazújváros-Hortobágy-Zámpuszta, co. H-B (Zoltai, 1907; 1910; Csalog, 1955; Korek, 1960; Kalicz and Makkay, 1977)	Middle
*27.	Balsa-Fecskepart, co. Szabolcs (Csallány and Gombás, 1965; Kalicz and Makkay, 1977)	Tiszadob Bükk I
*28.	Balsa-Vasuti hid, co. Szabolcs (Kalicz and Makkay, 1977)	Middle
*29.	Battonya-Gödrösök, co. Békés (Kalicz and Makkay, 1977)	Szakálhát
*30.	Békés-Povád, co. Békés ((Kalicz and Makkay, 1977)	Szakálhát
31.	Békésszentandrás, co. Békés (Kalicz and Makkay, 1977)	
*32.	Békésszentandrás-Körösön túli része, co. Békés (Kalicz and Makkay, 1977, same as Szarvas I)	Szakálhát
*33.	Bekölce, co. Heves (Korek and Patay, 1958; Kalicz, 1969; Kalicz and Makkay, 1977)	Tiszadob, Bükk

34.	Bélapatfalva, co. Heves (Korek and Patay, 1958; Kalicz, 1969; Kalicz and Makkay, 1977)	
*35.	Berekböszörmény-Korhányhalom, co. H-B (Mesterházy, 1966; Kalicz and Makkay, 1977)	Esztár
*36.	Berettyoújfalu-Malom, co. Hajdu-Bihar (Kalicz and Makkay, 1977)	Middle
*37.	Berettyoújfalu-Déli határrész, co. Hajdu-Bihar (Kalicz and Makkay, 1977)	Early
*38.	Biharkeresztes-Atand-Nagyfarkasdomb, co. H-B (Mesterházy, 1966; 1966a; Kalicz and Makkay, 1977)	Esztár
*39.	Biharkeresztes-Vasutáilomás, co. H-B (Kalicz and Makkay, 1977)	Middle, Esztár
40.	Bodrogkeresztúr-környéke, co. Borsod (Kalicz and Makkay, 1977)	
*41.	Bodrogkeresztúr-Széchényi- Wolkenstein kert, co. B. (Patay, 1961-; Kalicz and Makkay, 1977)	Early, Late AVK, Tiszadob, Bükk
*42.	Bodrogkeresztúr-Téglagyár, co. Borsod (Kalicz and Makkay, 1977)	Late AVK, Tiszadob
*43.	Bogács-Hintóvölgy, co. Borsod (Korek, 1960; Kalicz and Makkay, 1977)	Late AVK
*44.	Bogács-Kútivölgy, co. Borsod (Korek, 1960; Kalicz and Makkay, 1977)	Early
*45.	Bogács-Pazsagpuszta, co. Borsod (Korek, 1960; Kalicz and Makkay, 1977)	Szakálhát, Szilmeg
*46.	Bokros-Buzásdülö, co. Csongrád (Kalicz and Makkay, 1977)	Early
*47.	Boldogköváralja-Leányvár, co. Borsod (Mihalik, 1897; Tompa, 1929; Korek and Patay, 1958; Saás, 1959; Vértes, 1965; Kalicz and Makkay, 1977)	Bükk III
*48.	Boldogköváralja-Tekeres patak, co. Borsod (Kemenczei and Végh, 1964; Kemenczei, 1964; Kalicz and Makkay, 1977)	Late AVK, Tiszadob, Bükk I
*49.	Boldva, co. Borsod (Korek and Patay, 1958; Kalicz and Makkay, 1977)	
*50.	Borsodbóta-Kiskapudi vasutállomás, co. Borsod (Korek and Patay, 1958; Kalicz and Makkay, 1977)	
*51.	Borsodbóta-Rákóczi u. 6, co. Borsod (Kemenczei, 1969; Kemenczei and Végh, 1970)	Late AVK, Bükk I

52. Buj-Csépány tanya, co. Szabolcs
(Tompa, 1929; Korek and Patay, 1958;
Kalicz and Makkay, 1977)

53. Buj-Kiscsere, co. Szabolcs
(Kalicz and Makkay, 1977)

*54. Bükkaranyos-Földvár, co. Borsod — Szilmeg,
(Korek and Patay, 1958; Kemenczei, 1966; — Esztár,
Kemenczei and Végh, 1966; Kalicz and Makkay, 1977) — Bükk III

*55. Bükkszentlászló-Óhuta-Nagysánc, co. Borsod — Tiszadob
(Korek and Patay, 1958; Kalicz and Makkay, 1977)

*56. Csépa-Csipsárpart I, co. Szolnok — Szakálhát
(Kalicz, 1957; Kalicz and Makkay, 1977)

*57. Csépa-Csipsárpart II, co. Szolnok — Szakálhát
(Kalicz, 1957; Kalicz and Makkay, 1977)

*58. Csépa-Csipsárpart VI, co. Szolnok — Szakálhát
(Kalicz, 1957; Kalicz and Makkay, 1977)

59. Csépa-Csipsárpart VIII, co. Szolnok
(Kalicz and Makkay, 1977)

*60. Csépa-Ludasdülö- Szélmalom, co. Szolnok — Szakálhát
(Kalicz, 1957; Kalicz and Makkay, 1977)

*61. Csépa-Melegér, co. Szolnok — Szakálhát
(Kalicz, 1957; Kalicz and Makkay, 1977)

*62. Csépa-Téglaház, co. Szolnok — Szakálhát
(Kalicz, 1957; Kalicz and Makkay, 1977)

*63. Csépa-Telekpart, co. Szolnok — Szakálhát
(Kalicz, 1957; Kalicz and Makkay, 1977)

*64. Cserépfalu-Füzerkö barlang, co. Borsod — Bükk III
(Kadic, 1933; Korek and Patay, 1958;
Kalicz and Makkay, 1977)

*65. Cserépfalu-Hórvölgy, co. Borsod — Tiszadob
(Kalicz and Makkay, 1977)

66. Cserépfalu-Mészárnyék, co. Borsod
(Korek and Patay, 1958; Kalicz and Makkay, 1977)

*67. Csokvaomány, co. Borsod
(Korek and Patay, 1958; Kalicz and Makkay, 1977)

68. Csongrád-Túri, co. Csongrád
(Kalicz and Makkay, 1977)

*69. Debrecen-Lóversenytér, co. Hajdu-Bihar — Esztár,
(Kalicz and Makkay, 1977) — Szilmeg

*70. Debrecen-Nyulas-Transzformátor telep, co. H-B — Szilmeg
(Korek, 1960; Kalicz and Makkay, 1977)

71. Debrecen-Szabocs u. 4, co. Hajdu-Bihar
(Korek, 1960; Kalicz and Makkay, 1977)

*72. Debrecen-Szepesfalva, co. Hajdu-Bihar — Esztár
(Kalicz and Makkay, 1977)

*73. Debrecen-Tócóvölgy-Lászlóhalom, co. Hajdu-Bihar — Esztár, Szilmeg
(Zoltai, 1927; 1933; Tompa, 1929; Korek and Patay, 1956; Kalicz and Makkay, 1977)

*74. Debrecen-Tócóvölgy-Szántay halom, co. H-B — Esztár, Szilmeg
(Zoltai, 1927; Korek and Patay, 1956; Kalicz and Makkay, 1977)

*75. Dédestapolcsány-Gerennavár, co. Borsod — Middle
(Korek and Patay, 1958; Kalicz and Makkay, 1977)

*76. Demecser-Borzsava puszta, co. Szabolcs — Middle
(Korek, 1960; Kalicz and Makkay, 1977)

*77. Demecser-Vasuti örháza, co. Szabolcs — Middle
(Kalicz and Makkay, 1977)

78. Demjén, co. Heves
(Kalicz, 1969)

*79. Derecske-Bikás, co. Hajdu-Bihar — Middle
(Kalicz and Makkay, 1977)

*80. Derecske-Lyukashalom I, co. Hajdu-Bihar — Late AVK
(Kalicz and Makkay, 1977)

*81. Derecske-Lyukashalom II, co. Hajdu-Bihar — Late AVK
(Kalicz and Makkay, 1977)

*82. Derecske-Molnárdomb, co. Hajdu-Bihar — Middle
(Kalicz and Makkay, 1977)

*83. Derecske-Nagyszölö-Kállópart, co. Hajdu-Bihar — Middle
(Kalicz and Makkay, 1977)

*84. Derecske-Nagytanya I, co. Hajdu-Bihar — Middle
(Kalicz and Makkay, 1977)

*85. Derecske-Nagytanya II, co. Hajdu-Bihar — Middle
(Kalicz and Makkay, 1977)

*86. Derecske-Nyalakodó, co. Hajdu-Bihar — Middle
(Kalicz and Makkay, 1977)

*87. Derecske-Palocsa, co. Hajdu-Bihar — Middle
(Kalicz and Makkay, 1977)

*88. Derecske-Sámsoni domb, co. Hajdu-Bihar — Middle
(Kalicz and Makkay, 1977)

*89. Derecske-Téglagyár, co. Hajdu-Bihar — Esztár, Szilmeg
(Kalicz and Makkay, 1977)

90.	Deszk A, co. Csongrád (Kalicz and Makkay, 1977)	Szilmeg
*91.	Dévaványa-Atyaszeg-Hajashalom, co. Békés (Berczki, 1958; Kalicz and Makkay, 1977)	Szakálhát
*92.	Dévaványa-Bikahalom, co. Békés (Kalicz and Makkay, 1977)	Early
*93.	Dévaványa-Borszeg I, co. Békés (Horváth, 1959; Kalicz and Makkay, 1977)	Early
*94.	Dévaványa-Borszeg II, co. Békés (Horvath, 1959; Kalicz and Makkay, 1977)	
*95.	Dévaványa-Külső-Atyaszeg, co. Békés (Kalicz and Makkay, 1977)	Early
*96.	Dévaványa-Réheli gát, co. Békés (Gazdapusztai, 1959; Kalicz and Makkay, 1977)	Late AVK, Szilmeg, Szakálhát
*97.	Dévaványa-Sártó, co. Békés (Korek, 1960; 1961; Kalicz and Makkay, 1977)	Szilmeg, Szakálhát
*98.	Dévaványa-Simasziget, co. Békés (Korek, 1963; Kalicz and Makkay, 1977)	Szilmeg, Szakálhát
*99.	Ebes-Vérvölgy, co. Hajdu-Bihar (Mesterházy, 1966; Nepper and Máthė, 1971; Kalicz and Makkay, 1977)	Szatmár II
*100.	Edelény-Borsod-Derekegyháza, co. Borsod (Tompa, 1937; Csalog, 1955; Korek and Patay, 1958; Vértes, 1965; Kalicz and Makkay, 1977)	Bükk II, III
*101.	Edelény-Finke-Főutca, co. Borsod (Korek and Patay, 1958; Kalicz and Makkay, 1977)	Bükk I
*102.	Edelény-Kőbánya, co. Borsod (Korek and Patay, 1958; Kalicz and Makkay, 1977)	Late AVK, Tiszadob
*103.	Eger-Eger Szálloda, co. Heves (Kalicz and Makkay, 1977)	Middle
*104.	Eger-Kiseged-Egedi dülő, co. Heves (Galván, 1958; Korek and Patay, 1958; Kalicz, 1969; Kalicz and Makkay, 1977)	Szilmeg, Bükk III
*105.	Eger-Klapka Gy. u., co. Heves (Szabó, 1962; Kalicz and Makkay, 1977)	Middle
106.	Eger-Maklyán vár, co. Heves (Kalicz and Makkay, 1977)	
107.	Egyek-Rózsástelek, co. Hajdu-Bihar (Korek, 1957; 1960; 1961; JKalicz and Makkay, 1977)	
*108.	Emőd-Nagyhalom, co. Borsod (Korek and Patay, 1958; Kalicz and Makkay, 1977)	Tiszadob

*109.	Encs-Kastélydomb, co. Borsod (Kalicz and Makkay, 1977)	Late AVK, Tiszadob
110.	Endrőd-Csejt-Kőhalom, co. Békés (Kalicz and Makkay, 1977)	
*111.	Endrőd-Nagylaposi Népház, co. Békés (Banner, 1942; Korek, 1960; Kalicz and Makkay, 1977)	Szakálhát
*112.	Esztár-Hullámtér-Kiserdő, co. Hajdu-Bihar (Kutzián, 1963; Kalicz and Makkay, 1977)	Esztár
*113.	Fancsal-Hriczu-János kertje, co. Borsod (Kalicz and Makkay, 1977)	Tiszadob, Bükk
*114.	Fancsal-Kovács műhely, co. Borsod (Kalicz and Makkay, 1977)	Tiszadob
*115.	Fancsal-Oszőlőhegy, co. Borsod (Kalicz and Makkay, 1977)	Tiszadob, Bükk I
*116.	Fancsal-Patakmeder, co. Borsod (Kalicz and Makkay, 1977)	Late AVK, Tiszadob
*117.	Fancsal-Patakmeder alsó, co. Borsod (Kalicz and Makkay, 1977)	Late AVK
*118.	Fancsal-Vadkútalja, co. Borsod (Kalicz and Makkay, 1977)	Tiszadob
119.	Felsőtárkány-Barátrét, co. Heves (Korek and Patay, 1958; Kalicz, 1969)	
*120.	Felsőtárkány-Petényi barlang, co. Heves (Hermann et al., 1956; Korek and Patay, 1958; Kalicz and Makkay, 1977)	Late AVK
*121.	Felsőtárkány-Várhegy, co. Heves (Párducz, 1963; 1964; 1970; Kalicz, 1969; Kalicz and Makkay, 1977)	Bükk
*122.	Fulókércs-Fáy dűlő, co. Borsod (Kalicz and Makkay, 1977)	Middle, Tiszadob, Bükk
*123.	Füzesabony-Kettőshalom 2sz. homokd., co. Heves (Szabó, 1962; Kalicz and Makkay, 1977)	Late AVK
*124.	Füzesabony-Pusztaszikszó, co. Heves (Kalicz, 1969; Kalicz and Makkay, 1977)	Szakálhát
*125.	Gáborján-Lófogópart, co. Hajdu-Bihar (Kalicz and Makkay, 1977)	Middle, Esztár
*126.	Gáborján-Vár, co. Hajdu-Bihar (Kalicz and Makkay, 1977)	Middle
*127.	Gáva-Dombsziget, co. Szabolcs (Kalicz and Makkay, 1977)	Middle, Bükk

*128.	Gáva-Katóhalom, co. Szabolc. (Jósa, 1915; Tompa, 1929; Korek and Patay, 1958; Kalicz and Makkay, 1977)	Late AVK, Tiszadob, Bükk I
*129.	Gelej-Kanális dülő, co. Borsod (Kalicz and Makkay, 1977)	Middle
*130.	Gelej-Ujszőlők, co. Borsod (Kalicz and Makkay, 1977)	Middle
*131.	Golop-Vizmosás, co. Borsod (Korek and Patay, 1958; Kalicz and Makkay, 1977)	
132.	Gyoma-Pőhalom, co. Békés (Kalicz and Makkay, 1977)	
133.	Gyöngyöspata-Előmáj, co. Heves (Kalicz and Makkay, 1977)	
*134.	Gyöngyöspata-Várhegy, co. Heves (Kozák, 1964)	Bükk
*135.	Hajduböszörmény-Köveshalom, co. Hajdu-Bihar (Nepper and Máthé, 1971)	Middle
*136.	Hajdunánás-Nagy Vidi halom, co. Hajdu-Bihar (Nepper and Máthé, 1971)	Middle
137.	Hajdusámson-Bajinka, co. Hajdu-Bihar (Kalicz and Makkay, 1977)	
138.	Hajdusámson-Majorsági földek, co. Hajdu-Bihar (Kalicz and Makkay, 1977)	
139.	Hajdusámson-Vermeshát, co. Hajdu-Bihar (Kalicz and Makkay, 1977)	
*140.	Hajduszóboszló-Téglagyár, co. Hajdu-Bihar (Kralovánszky, 1958; Korek, 1960; 1961 Kalicz and Makkay, 1977)	Middle
*141.	Halmaj-Vasonca patak, co. Borsod (Kemenczei, 1965; Kalicz and Makkay, 1977)	Late AVK, Bükk II
142.	Hejce, co. Borsod (Korek and Patay, 1958; Kalicz and Makkay, 1977)	
143.	Hejőkeresztúr I, co. Borsod (Korek and Patay, 1958; Kalicz and Makkay, 1977)	
144.	Hejőkeresztúr and Mezőcsát között, co. Borsod (Korek and Patay, 1958; Kalicz and Makkay, 1977)	
*145.	Hejőkeresztúr-Templomkert, co. Borsod (Kalicz and Makkay, 1977)	Bükk
*146.	Hejőkeresztúr, near Oszlár, co. Borsod) (Kalicz and Makkay, 1977)	Middle

*147. Hejőszalonta-Kavicsgödör, co. Borsod (Korek and Patay, 1958; Kalicz and Makkay, 1977)	Early
*148. Hencia-Csikostó, co. Hajdu-Bihar (Kalicz and Makkay, 1977)	Esztár
149. Hernádbüd, co. Borsod (Korek and Patay, 1958; Kalicz and Makkay, 1977)	
*150. Hernádcéce, co. Borsod (Megay, 1960; Kalicz and Makkay, 1977)	Bükk
*151. Hernádkak, co. Borsod (Korek and Patay, 1958; Korek, 1960)	Bükk
*152. Hernádnémeti-Bőcs-Vasutállomás, co. Borsod (Kemenczei, 1969; Kemenczei and Végh, 1970; Kalicz and Makkay, 1977)	Early
*153. Hernádnémeti-Vasutállomás, co. Borsod (Gnandt, 1957; Korek and Patay, 1958; Kalicz and Makkay, 1977)	Late AVK, Tiszadob
*154. Hidasnémeti-Vasutállomás, co. Borsod (Kemenczei and Végh, 1968; Kalicz and Makkay, 1977)	Late AVK, Tiszadob, Bükk I
*155. Hódmezővásárhely-Barci rét, co. Csongrád (Banner, 1940; Kalicz and Makkay, 1977)	Middle
156. Hódmezővásárhely-Gorzsa-Kovács, co. Csongrád (Gazdapusztai, 1957; Kalicz and Makkay, 1977)	
*157. Hódmezővásárhely-Kökénydomb, co. Csongrád (Banner, 1931a; 1932; 1945; 1951; Banner and Korek, 1949; Kalicz and Makkay, 1977)	Szakálhát
*158. Hódmezővásárhely-Szakálhát, co. Csongrád (Banner and Balint, 1935; Korek, 1960; Kalicz and Makkay, 1977)	Szakálhát
159. Hollóháza-Fagyosforrás, co. Borsod (Kemenczei and Végh, 1969; Kalicz and Makkay, 1977)	
*160. Hortobágy-Bivalyhalom, co. Hajdu-Bihar (Kalicz and Makkay, 1977)	Middle, Szilmeg
*161. Hortobágy-Csárda, co. Hajdu-Bihar (Kalicz and Makkay, 1977)	Middle
*162. Hortobágy-2 sz. Halászlaponyag, co. Hajdu-Bihar (Kalicz and Makkay, 1977)	Late AVK, Esztár
*163. Ibrány-Bleier birtok, co. Szabolcs (Korek, 1960; Kalicz and Makkay, 1977)	Szatmár
164. Ibrány-Bodzásdülő I, co. Szabolcs (Kalicz and Makkay, 1977)	

165. Ibrány-Bodzásdülő II, co. Szabolcs
(Kalicz and Makkay, 1977)

166. Ibrány-Dergitag-Észak, co. Szabolcs
(Kalicz and Makkay, 1977)

167. Ibrány-Kéthegydülő, co. Szabolcs
(Kalicz and Makkay, 1977)

168. Ibrány-Kismacskás, co. Szabolcs
(Kalicz and Makkay, 1977)

169. Ibrány-Nádastábla, co. Szabolcs
(Kalicz and Makkay, 1977)

170. Ibrány-Nagytanyasi tag, co. Szabolcs
(Kalicz and Makkay, 1977)

171. Ibrány-Szennai domb, co. Szabolcs
(Kalicz and Makkay, 1977)

172. Ibrány-Szőlőhomok, I, co. Szabolcs
(Kalicz and Makkay, 1977)

173. Jászalsószentgyörgy, co. Szolnok
(Kalicz and Makkay, 1977)

174. Jászberény-Cserőhalom, co. Szolnok
(Kalicz and Makkay, 1977)

175. Jászberény-Disznószög, co. Szolnok
(Kalicz and Makkay, 1977)

176. Jászberény-2 sz. gyár, co. Szolnok
(Kalicz and Makkay, 1977)

*177. Jászdózsa-Kápolna halom, co. Szolnok — Middle
(Korek, 1960; Kalicz and Makkay, 1977)

178. Jásztelek, co. Szolnok
(Kalicz and Makkay, 1977)

179. Jéke, co. Szabolcs
(Kalicz and Makkay, 1977)

180. Karcag, co. Szolnok
(Korek, 1968; Kalicz and Makkay, 1977)

*181. Karcag-Berekhalom, co. Szolnok
(Korek, 1960; Kalicz and Makkay, 1977)

182. Karcsa-TSZ szőlő, co. Borsod
(Kalicz and Makkay, 1977)

183. Kazincbarcika, co. Borsod
(Korek and Patay, 1958; Kalicz and Makkay, 1977)

*184. Kemecse-Vastanya, co. Szabolcs — Bükk
(Kalicz, 1958a; Kalicz and Makkay, 1977)

*185. Kenderes-Kulis, co. Szolnok (Kalicz and Makkay, 1977) — Late AVK

*186. Kenderes-Telekhalom. co. Szolnok (Kalicz and Makkay, 1977) — Middle

187. Kenézlő-Baji homok, co. Borsod (Kalicz and Makkay, 1977)

*188. Kenézlő-Fazekaszug, co. Borsod (Tompa, 1929; Kiss, 1939; Korek and Patay, 1958; Kutzián, 1963; Kalicz and Makkay, 1977) — Tiszadob, Bükk

*189. Kenézlő-Gicepart, co. Borsod (Kalicz and Makkay, 1977) — Middle

*190. Kenézlő-Kastélykert, co. Borsod (Kalicz and Makkay, 1977) — Tiszadob, Bükk

191. Kenézlő-Szérűskert, co. Borsod (Kiss, 1939; Patay, 1957; Korek and Patay, 1958; Kalicz and Makkay, 1977)

192. Kengyel-Csőbrősérpart, co. Szolnok (Kalicz and Makkay, 1977)

*193. Kengyel-Csőbrősérpart-Vágyogvető, co. Szolnok (Korek, 1960; Kalicz and Makkay, 1977) — Szakálhát

*194. Kengyel-Temető, co. Szolnok (Kalicz and Makkay, 1977) — Middle

*195. Kerecsend-Fácános, co. Heves (Kalicz and Makkay, 1977) — Late AVK

*196. Kétegyháza-Kétegyházi tanyák, co. Békés (Gazdapusztai, 1967; Kalicz and Makkay, 1977)

*197. Kisköre-Gát, co. Szolnok (Korek, 1965; 1967; 1968; Kalicz and Makkay, 1977) — Early

*198. Kisköre-Rákhát, co. Szolnok (Kalicz and Makkay, 1977) — Late AVK

199. Kismarja, co. Hajdu-Bihar (Kalicz and Makkay, 1977)

200. Kistokaj, co. Borsod (Korek and Patay, 1958; Kalicz and Makkay, 1977)

201. Kistokaj-Gerenda dűlő, co. Borsod (Korek and Patay, 1958; Kalicz and Makkay, 1977)

202. Kistokaj-Kistokaj tanya, co. Borsod (Korek and Patay, 1958; Kalicz and Makkay, 1977)

*203. Kisvárda-Darusziget, co. Szabolcs (Korek, 1960; Kalicz and Makkay, 1977) — Esztár

*204.	Kisvarsány-Tiszapart, co. Szabolcs (Korek, 1964; Kalicz and Makkay, 1977)	Esztár
205.	Kölcse, co. Szabolcs (Kalicz and Makkay, 1977)	
*206.	Konyár-Papfeneke, co. Hajdu-Bihar (Korek and Patay, 1956; Korek, 1960; Kalicz and Makkay, 1977)	Esztár
*207.	Konyár-Telekhatári gát, co. Hajdu-Bihar (Kalicz and Makkay, 1977)	Esztár
*208.	Konyár-Vasutmenti vályogvetök, co. Hajdu-Bihar (Kalicz and Makkay, 1977)	Esztár
*209.	Korlát-Fönya patak völgye, co. Borsod (Patay, 1941; Korek and Patay, 1958; Kalicz and Makkay, 1977)	Late AVK, Tiszadob, Bükk
210.	Körösladány, co. Békés (Korek, 1960; Kalicz and Makkay, 1977)	
211.	Köröstarcsa, co. Békés (Korek, 1960; Kalicz and Makkay, 1977)	
212.	Kotaj-Kertekalja, co. Szabolcs (Kalicz and Makkay, 1977)	
*213.	Kunszentmárton-Bohonya III, co. Szolnok (Kalicz, 1957; Kalicz and Makkay, 1977)	Late AVK, Szilmeg, Szakálhát
*214.	Kunszentmárton-Érpart I, co. Szolnok (Kalicz, 1954; Kalicz and Makkay, 1977)	Szakálhát
*215.	Kunszentmárton-Érpart II, co. Szolnok (Kalicz, 1954; Kalicz and Makkay, 1977)	Szakálhát
*216.	Kunszentmárton-Érpart III, co. Szolnok (Kalicz, 1954; Kalicz and Makkay, 1977)	Middle
*217.	Kunszentmárton-Érpart IV, co. Szolnok (Kalicz, 1954; Kalicz and Makkay, 1977)	Middle
*218.	Kunszentmárton-Érpart V, co. Szolnok (Kalicz, 1954; Kalicz and Makkay, 1977)	Szakálhát
*219.	Kunszentmárton-Érpart VI, co. Szolnok (Kalicz, 1954; Kalicz and Makkay, 1977)	Middle, Szakálhát
*220.	Kunszentmárton-Érpart VII, co. Szolnok (Kalicz, 1954; Kalicz and Makkay, 1977)	Late AVK, Szakálhát
*221.	Kunszentmárton-Érpart VIII-Bábocka II, co. Sz. (Kalicz, 1954; Kalicz and Makkay, 1977)	Late AVK, Szakálhát
*222.	Kunszentmárton-Érpart IX, co. Szolnok (Kalicz, 1954; Kalicz and Makkay, 1977)	Late AVK, Szakálhát

*223. Kunszentmárton-Érpart-Kováshalom, co. Szolnok (Kalicz, 1954; Kalicz and Makkay, 1977) — Szakálhát

*224. Kunszentmárton-Érpart-40. lelethely, co. Szolnok (Kalicz, 1954; Kalicz and Makkay, 1977) — Szakálhát

*225. Kunszentmárton-Kékes II, co. Szolnok (Kalicz, 1957; Kalicz and Makkay , 1977) — Szakálhát

*226. Kunszentmárton-Kékes III-Téglaégető, co. Szolnok (Kalicz, 1957 ; Kalicz and Makkay, 1977) — Late AVK, Szakálhát

*227. Kunszentmárton-Kékes IV, co. Szolnok (Kalicz, 1957; Kalicz and Makkay, 1977) — Szakálhát

*228. Kunszentmárton-Kékes V, co. Szolnok (Kalicz, 1957; Kalicz and Makkay, 1977) — Szakálhát

*229. Kunszentmárton-Kettőshalom, co. Szolnok (Csallany, 1939; Kalicz and Makkay, 1977) — Szakálhát

*230. Kunszentmárton-Kökényzug II, co. Szolnok (Kalicz, 1954; Kalicz and Makkay, 1977) — Middle, Szakálhát

*231. Kunszentmárton-Kungyalu-Körtvélyes I, co. Sz. (Kalicz, 1957; Kalicz and Makkay, 1977) — Late AVK, Szakálhát

*232. Kunszentmárton-Kungyalu-Körtvélyes III, co. Sz. (Kalicz, 1957; Kalicz and Makkay, 1977) — Szakálhát

*233. Kunszentmárton-Kungyalu-Körtvélyes IV, co. Sz. (Kalicz, 1957; Kalicz and Makkay, 1977) — Middle

*234. Kunszentmárton-Kungyalu-Körtvélyes VI, co. Sz. (Kalicz, 1957; Kalicz and Makkay, 1977) — Middle

235. Legyesbénye, co. Borsod (Korek and Patay, 1958; Kalicz and Makkay, 1977)

236. Legyesbenye-Futóbarlang, co. Borsod (Korek and Patay, 1958; Kalicz and Makkay, 1977)

*237. Leninváros-Lakkfestékgyár, co. Borsod (Kalicz and Makkay, 1977) — Middle

238. Mád-Galambos dűlő, co. Borsod (Kalicz and Makkay, 1977)

239. Mályinka, co. Borsod (Korek, 1957; Korek and Patay, 1958; Kalicz and Makkay, 1977)

240. Martonyi Barlang, co. Borsod (Korek and Patay, 1958; Kalicz and Makkay, 1977)

*241. Megyaszó-Csákó, co. Borsod (Tompa, 1937; Korek and Patay, 1958; Kalicz and Makkay, 1977) — Late AVK, Tiszadob

*242. Megyaszó-Komáromi domb, co. Borsod (Korek and Patay, 1958; Kalicz and Makkay, 1977)

*243. Megyaszó-Majoros, co. Borsod (Tompa, 1937; Kalicz and Makkay, 1977)	Tiszadob
244. Meszes-Kigyószög, co. Borsod (Korek and Patay, 1958; Kalicz and Makkay, 1977)	
*245. Mezőcsát-Eperjes tanya, co. Borsod (Kalicz and Makkay, 1977)	Middle
246. Mezőcsát-Erdőigazgatoság csemete kertje, co. Bo. (Kalicz and Makkay, 1977)	
247. Mezőcsát-Harangdomb, co. Borsod (Kalicz and Makkay, 1977)	
*248. Mezőcsát-Hőrcsögös, co. Borsod (Castiglione, 1970; Kalicz and Makkay, 1977)	Middle
*249. Mezőcsát-Keselyűhalom I-Homokbánya, co. Borsod (Kalicz and Makkay, 1977)	Middle
*250. Mezőcsát-Keselyűhalom II-Kavicsbánya, co. Borsod (Kalicz and Makkay, 1977)	Middle
*251. Mezőcsát-Kiszeli tanya, co. Borsod (Kalicz and Makkay, 1977)	Middle
*252. Mezőcsát-Nagyhatárdomb, co. Borsod (Kalicz and Makkay, 1977)	Middle
*253. Mezőcsát-Pásthy domb, co. Borsod (Korek and Patay, 1958; Kalicz and Makkay, 1977)	
*254. Mezőkeresztes-Hegyeshalom, co. Borsod (Korek and Patay, 1958; Kalicz and Makkay, 1977)	Late AVK, Szilmeg
*255. Mezőkeresztes-Lengyel tanya, co. Borsod (Korek and Patay, 1958; Korek, 1960; Kalicz and Makkay, 1977)	Early
256. Mezőkövesd-Déli határrész, co. Borsod (Kemenczei and Végh, 1971; Kalicz and Makkay, 1977)	
*257. Mezőnagymihály-Baglyasdülő, co. Borsod (Kalicz and Makkay, 1977)	Middle
*258. Mezőnyarád-Vörösmarty u. vége, co. Borsod (Megay, 1962; Kalicz and Makkay, 1977)	Middle
*259. Mezőpeterd-Széchényi u., co. Hajdu-Bihar (Nepper and Máthé, 1971)	Esztár
260. Mezőtárkány-Halastó környéke, co. Borsod (Korek, 1960; Kalicz and Makkay, 1977)	
*261. Mezőtúr-Berettyópart, co. Szolnok (Kalicz and Makkay, 1977)	Middle, Szakálhát
*262. Mezőtúr-Berettyópart-Vasuti hid, co. Szolnok (Kalicz and Makkay, 1977)	Late AVK

263.	Mezötúr-Szántó, co. Szolnok (Kalicz and Makkay, 1977)	
264.	Mezötúr határa, co. Szolnok (Kalicz and Makkay, 1977)	
*265.	Mikóháza, co. Borsod (Kemenczei and Végh, 1969; Kalicz and Makkay, 1977)	Tiszadob
*266.	Miskolc-Büdöspest barlang, co. Borsod (Kadic, 1916; Tompa, 1929; 1937; Korek and Patay, 1958; Kalicz and Makkay, 1977)	Late AVK, Tiszadob, Bükk I, II
267.	Miskolc-Diósgyöri Tapolca barlang, co. Borsod (Szendrei, 1883; Saád and Gaál, 1934; Korek and Patay, 1958; Kalicz and Makkay, 1977)	
*268.	Miskolc-Fütöház, co. Borsod (Korek and Patay, 1958; Korek, 1959a; 1960; 1970; Kalicz and Makkay, 1977)	Middle, Late AVK, Tiszadob, Bükk I
*269.	Miskolc-Háromkuti barlang, co. Borsod (Kadic, 1914; Korek and Patay, 1958; Kalicz and Makkay, 1977)	Late AVK, Bükk
270.	Miskolc-Hejöcsaba, co. Borsod (Korek and Patay, 1958; Tompa, 1929; Kalicz and Makkay, 1977)	
*271.	Miskolc-Herman Ottó barlang and Köfülke, co. Bo. (Kadic, 1916; 1934; Tompa, 1929; 1937; Korek and Patay, 1958; Kalicz and Makkay, 1977)	Bükk
*272.	Miskolc-Hillebrand barlang, co. Borsod (Párducz, 1949; Korek, 1957; 1958; Kalicz and Makkay, 1977)	Late AVK, Tiszadob, Szilmeg, Bükk
*273.	Miskolc-Kecskelyuk barlang, co. Borsod (Kadic, 1907; 1934; Korek and Patay, 1958; Kalicz and Makkay, 1977)	Bükk
*274.	Miskolc-Kinizsi u., co. Borsod (Korek and Patay, 1958; Kalicz and Makkay, 1977)	Late AVK, Tiszadob
275.	Miskolc-Létrástetö-Túskos barlang, co. Borsod (Kemenczei and Végh, 1971; Kalicz and Makkay, 1977)	
276.	Miskolc-Puskaporos köfülke, co. Borsod (Kadic, 1907; Korek and Patay, 1968; Kalicz and Makkay, 1977)	
*277.	Miskolc-Repülötér, co. Borsod (Kemenczei and Végh, 1966; 1968; Korek and Patay, 1958; Kalicz and Makkay, 1977)	Late AVK, Tiszadob
*278.	Miskolc-Szeleta barlang, co. Borsod (Saád, 1929; Tompa, 1929; Kalicz and Makkay, 1977)	Late AVK, Bükk

*279. Miskolc-Szirma-Fáskert, co. Borsod (Korek and Patay, 1958; Kalicz and Makkay, 1977)	Middle, Tiszadob
*280. Miskolc-Szirma-Sajópart, co. Borsod (Korek and Patay, 1958; Kalicz and Makkay, 1977)	Middle
*281. Miskolc-Tapolca-Szentkereszthelyi barlang, co. B (Korek and Patay, 1958; Kalicz and Makkay, 1977)	Bükk
*282. Miskolc-Tapolca-Várhegy, co. Borsod (Kalicz and Makkay, 1977)	Middle, Szilmeg
283. Miskolc-Tapolca-Vizesbarlang, co. Borsod (Korek and Patay, 1958; Kalicz and Makkay, 1977)	
*284. Miskolc-Tüzér u., co. Borsod (Kalicz and Makkay, 1977)	Middle
*285. Miskolc-Vidróczky barlang, co. Borsod (Korek and Patay, 1958; Kalicz and Makkay, 1977)	Bükk I
*286. Miskolc-Zsarnyay tanya, co. Borsod (Korek and Patay, 1958; Kalicz and Makkay, 1977)	Early, Late AVK, Bükk
*287. Muhi-Kocsmadomb, co. Borsod (Korek and Patay, 1958; Kalicz and Makkay, 1977)	Middle
*288. Nádudvar-Csukáskert, co. Hajdu-Bihar (Korek, 1960; Kalicz and Makkay, 1977)	Middle
289. Nádudvar-Illetőségi föld, co. Hajdu-Bihar (Korek, 1960; Kalicz and Makkay, 1977)	
*290. Nagyecsed-Péterzug, co. Szabolcs (Kalicz and Makkay, 1972a; 1977)	Szatmár I
*291. Nagyecsed-Ráckóczi vár, co. Szabolcs (Kalicz and Makkay, 1977)	Middle, Late
292. Nagyhalász-Málèhegy, co. Szabolcs (Kalicz and Makkay, 1977)	
*293. Nagykálló-Strandfürdő, co. Szabolcs (Korek, 1957; Korek and Patay, 1958; Kalicz and Makkay, 1977)	Late AVK, Tiszadob, Esztár
*294. Nagykörü-Öregkarád dülö, co. Szolnok (Kalicz and Makkay, 1977)	Late AVK, Szakálhát
*295. Nagykörü-Tiszapart-Hidashát, co. Szolnok (Kalicz and Makkay, 1977)	Szakálhát
*296. Nagyrév-Sápipart I, co. Szolnok (Kalicz, 1957; Kalicz and Makkay, 1977)	Szakálhát
*297. Nagyrév-Sápipart II, co. Szolnok (Kalicz, 1957; Kalicz and Makkay, 1977)	Szakálhát

*298.	Nagyrév-Sápipart III, co. Szolnok (Kalicz, 1957; Kalicz and Makkay, 1977)	Szakálhát
*299.	Nagyrév-Temetőpart, co. Szolnok (Kalicz, 1957; Kalicz and Makkay, 1977)	Middle
*300.	Nagyút-Danka, co. Heves (Kalicz and Makkay, 1977)	Szakálhát
301.	Nagyvisnyó-Olyvesköi rókalyuk, co. Borsod (Korek and Patay, 1958; Kalicz and Makkay, 1977)	
*302.	Nyiregyháza-Bujtos-Morgo-Temető, co. Szabolcs (Korek and Patay, 1958; Kalicz and Makkay, 1977)	Late AVK, Tiszadob, Esztár
303.	Nyiregyháza-Csikókarám, co. Szabolcs (Kalicz and Makkay, 1977)	
304.	Nyíri, co. Borsod (Kalicz and Makkay, 1977)	
305.	Nyirpazony-Igrice, co. Szbolcs (Kalicz and Makkay, 1977)	
*306.	Ocsöd-Daruvásárhát, co. Szolnok (Kalicz and Makkay, 1977)	Late AVK
*307.	Ocsöd-Harangdülö, co. Szolnok (Kalicz and Makkay, 1977)	Late AVK
*308.	Ocsöd Hidfö I, co. Szolnok (Schupiter, 1931; Kalicz, 1957; Kalicz and Makkay, 1977)	Early
*309.	Ocsöd Hidfö II, co. Szolnok (Kalicz and Makkay, 1977)	Middle
*310.	Ocsöd-Kenderhalom, co. Szolnok (Csalog, 1959; Kalicz and Makkay, 1977)	Middle
311.	Ocsöd-Kéthalom, co. Szolnok (Korek, 1959; 1961; Kalicz and Makkay, 1977)	
*312.	Onga-Vasuti Őrház, co. Borsod (Korek, 1957; Korek and Patay, 1958; Kalicz and Makkay, 1977)	Tiszadob, Bükk I
*313.	Onod-Katolikus temető- co. Borsod (Korek and Patay, 1958; Kalicz and Makkay, 1977)	Tiszadob
*314.	Oros-4 sz. müút 7600-7767 m, co. Szabolcs (Korek, 1951; Korek and Patay, 1958; Kalicz and Makkay, 1977)	Tiszadob, Szamos
*315.	Oros-Földvár, co. Szabolcs (Erdélyi, 1965; Kalicz and Makkay, 1977)	Bükk
*316.	Ozd-Bolyok II. kerület (Kalicz and Makkay, 1977)	Late AVK, Tiszadob

*317.	Pálháza-Borzsava patak völgye, co. Borsod (Korek, 1969; Kalicz and Makkay, 1977)	Bükk I
*318.	Paszab-Eghalom, co. Szabolcs (Korek and Patay, 1958; Kalicz and Makkay, 1977)	Bükk II and III
319.	Paszab-Felnémet tag, co. Szabolcs (Kalicz and Makkay, 1972a; 1977)	
*320.	Paszab-Görömbey kert, co. Szabolcs (Kalicz and Makkay, 1977)	Esztár
*321.	Paszab-Hordozó dülö, co. Szabolcs (Tompa, 1929; 1937; Kalicz and Makkay, 1977)	Szilmeg
322.	Paszab-Medejdülö, co. Szabolcs (Korek and Patay, 1958; Kalicz and Makkay, 1977)	
*323.	Paszab-Öved, co. Szabolcs (Korek and Patay, 1958; Kalicz and Makkay, 1977)	Bükk
324.	Paszab-Szaniszló, co. Szabolcs (Kalicz and Makkay, 1977)	
325.	Paszab-Terebélyes, co. Szabolcs (Kalicz and Makkay, 1977)	
*326.	Paszab-Zádó, co. Szabolcs (Korek and Patay, 1958; Kalicz and Makkay, 1977)	Late AVK, Tiszadob, Bükk III
*327.	Pócspetri-Ercsi vár, co. Szabolcs (Nemeth, 1967; Kalicz and Makkay, 1977)	Esztár
*328.	Polgár-Basatanya, co. Hajdu-Bihar (Kutzián, 1963; Kalicz and Makkay, 1977)	Late AVK, Szilmeg, Bükk
*329.	Polgár-Csöszdomb, co. Hajdu-Bihar (Kutzián, 1958; Kalicz and Makkay, 1977)	Middle
*330.	Polgár-Folyás-Bivalyhalom, co. Hajdu-Bihar (Nepper and Máthé, 1971)	Szilmeg
*331.	Polgár-Folyás-Iskola, co. Hajdu-Bihar (Kalicz and Makkay, 1977)	Middle
*332.	Polgár-Folyás-Szilmeg, co. Hajdu-Bihar (Korek and Patay, 1958; Kutzián, 1963; Kalicz and Makkay, 1977)	Szilmeg, Esztár, Bükk
*333.	Polgár-Kenderföldek, co. Hajdu-Bihar (Nepper and Máthé, 1971; Kalicz and Makkay, 1977)	Late AVK, Tiszadob, Esztár, Bükk
334.	Polgár-Kistanya, co. Hajdu-Bihar (Korek, 1960; Kalicz and Makkay, 1977)	
335.	Polgár-98. magasági pont, co. Hajdu-Bihar (Nepper and Máthé, 1971)	

*336.	Poroszló-Ráboly puszta, co. Heves (Patay, 1969; Kalicz and Makkay, 1977)	Middle
*337.	Prügy-Tökföld, co. Borsod (Korek and Patay, 1958; Kemenczei and Végh, 1969; Kalicz and Makkay, 1977)	Late AVK, Tiszadob, Bükk
*338.	Püspökladány-Mérgeskut, co. Hajdu-Bihar (Kalicz and Makkay, 1977)	Esztár
*339.	Püspökladány-Ürmöshát, co. Hajdu-Bihar (Kalicz and Makkay, 1977)	Middle, Esztár
*340.	Rakamaz-Timár morotvaomlás, co. Szabolcs (Tompa, 1929; Korek and Patay, 1958; Kalicz and Makkay, 1977)	Late AVK, Tiszadob, Bükk
*341.	Rakamaz-Vasutállomás, co. Szabolcs (Kalicz and Makkay, 1977)	Middle
*342.	Rétközberencs-Liget, co. Szabolcs (Kalicz and Makkay, 1977)	Middle
*343.	Rétközberencs-Páromdomb, co. Szabolcs (Kalicz, 1956; 1958a; Kalicz and Makkay, 1972a, 1977)	Szatmár II
344.	Sajóecseg, co. Borsod (Tompa, 1929; Korek and Patay, 1958; Kalicz and Makkay, 1977)	
*345.	Sajókeresztur-Homokbánya, co. Borsod (Korek and Patay, 1958; Kalicz and Makkay, 1977)	Late AVK, Tiszadob, Bükk I
*346.	Sajókeresztur-Malom, co. Borsod (Korek and Patay, 1958; Kalicz and Makkay, 1977)	Bükk
*347.	Sajókeresztur III, co. Borsod (Korek and Patay, 1958; Kemenczei and Végh, 1970; Kalicz and Makkay, 1977)	Bükk
348.	Sajószentpeter, co. Borsod (Korek and Patay, 1958; Kalicz and Makkay, 1977)	
349.	Sály, co. Borsod (Korek and Patay, 1958; Kalicz and Makkay, 1977)	
*350.	Sáp-Temető és vasutállomás között, co. Borsod (Nepper and Máthé, 1971)	Szilmeg
*351.	Sárazsadány-Akasztószer, co. Borsod (Kutzián, 1959; 1970)	Bükk
*352.	Sárazsadány-Major, co. Borsod (Kalicz and Makkay, 1977)	Middle

*353.	Sárazsadány-Templomdomb, co. Borsod (Korek and Patay, 1958; Kutzián, 1959; 1970; Kalicz, 1958b; Kalicz and Makkay, 1977)	Middle, Late AVK, Tiszadob, Bükk
*354.	Sárospatak-Rákoczi vár, co. Borsod (Kovalovszki, 1964; Kalicz and Makkay, 1977)	Bükk
*355.	Sátoraljaujhely-Ronyva patak, co. Borsod (Visegrádi, 1907; 1912; Tompa, 1929; Korek and Patay, 1958; Kalicz and Makkay, 1977)	Tiszadob, Bükk, Sátor-aljaujhely
356.	Sima, co. Borsod (Kemenczei and Végh, 1971; Kalicz and Makkay, 1977)	
*357.	Szabolcs-Földvár, co. Szabolcs (Németh, 1972; Kalicz and Makkay, 1977)	Middle
*358.	Szakoly-Bornemissza K. földje, co. Szabolcs (Korek and Patay, 1958; Kalicz and Makkay, 1977)	Bükk III
*359.	Szamossályi-Szamospart, co. Szabolcs (Korek, 1968; Kalicz and Makkay, 1977)	Esztár
360.	Szászberek-Alsószászberek-Bogárzó, co. Szolnok (Korek, 1960; Kalicz and Makkay, 1977)	
*361.	Szászberek-Felsőszászberek-Berenta II, co. Szolnok (Korek, 1960; Kalicz and Makkay, 1977)	Szilmeg, Szakálhát
*362.	Szarvas-Érpart, co. Békés (Krecsmárik, 1915; Banner, 1931; Kalicz and Makkay, 1977)	Late AVK
*363.	Szarvas-Kovácshalom, co. Békés (Banner, 1931; Kalicz and Makkay, 1977)	Middle
*364.	Szarvas-Ótemető, co. Békés (Krecsmarik, 1915; Banner, 1931; Kalicz and Makkay, 1977)	Late AVK
365.	Szarvas-Szappanos, co. Békés (Banner, 1931; Kalicz and Makkay, 1977)	
*366.	Szarvas-Vasuti hid, co. Békés (Krecsmárik, 1911; Banner, 1931; Kalicz and Makkay, 1977)	Late AVK
367.	Szeged-Siróhalom, co. Csongrád (Kalicz and Makkay, 1977)	
*368.	Szeghalom-Dioőri csatorna, co. Békés (Ecsedy, 1972a; Kalicz and Makkay, 1977)	Szakálhát
*369.	Szeghalom-Kovács halom, co. Békés (Darnay, 1905; Szeghalmy, 1913; Bakay, 1971; Kalicz and Makkay, 1977)	Szilmeg, Szakálhát
*370.	Szegvár-Tűzköves, co. Csongrád (Schupiter, 1931; Csalog, 1958; 1965; Korek, 1971; Kalicz and Makkay, 1977)	Szakálhát

*371. Szelevény-Bivalyos, co. Szolnok (Kalicz, 1957; Kalicz and Makkay, 1977) — Szakálhát

*372. Szelevény-Hosszuhát, co. Szolnok (Kalicz, 1957; Kalicz and Makkay, 1977) — Szakálhát

*373. Szelevény-Kisasszonypart II, co. Szolnok (Kalicz, 1957; Kalicz and Makkay, 1977) — Szakálhát

*374. Szelevény-Menyasszonypart II, co. Szolnok (Kalicz, 1957; Kalicz and Makkay, 1977) — Middle

*375. Szelevény-Pecekpart I, co. Szolnok (Kalicz, 1957; Kalicz and Makkay, 1977) — Szakálhát

*376. Szelevény-Pecekpart II, co. Szolnok (Kalicz, 1957; Kalicz and Makkay, 1977) — Middle

*377. Szelevény-Telekpart I, co. Szolnok (Kalicz, 1957; Kalicz and Makkay, 1977) — Middle

*378. Szelevény-Telekpart II, co. Szolnok (Kalicz, 1957; Kalicz and Makkay, 1977) — Szakálhát

*379. Szelevény-Tóközipart-Vasuti átvágás, co. Szolnok (Kalicz, 1957; Kalicz and Makkay, 1977) — Szakálhát

380. Szelevény-Vadas, co. Szolnok (Kalicz and Makkay, 1977)

381. Szemere, co. Borsod (Korek and Patay, 1958; Kalicz and Makkay, 1977)

382. Szendrö-Ördöggáti Csengöbarlang, co. Borsod (Korek and Patay, 1958; Kalicz and Makkay, 1977)

*383. Szentes-Berekhát, co. Csongrád (Schupiter, 1931; Kalicz and Makkay, 1977) — Szakálhát

*384. Szentes-Bökény-Köröspart, co. Csongrád (Kovalovszki, 1957; Kalicz and Makkay, 1977) — Szakálhát

*385. Szentes-Érpart-Szentilona, co. Csongrád (Kovalovszki, 1957; Csalog, 1964; Kalicz and Makkay, 1977) — Middle, Szilmeg, Szakálhát

*386. Szentes-Jámborhalom, co. Csongrád (Korek, 1960; Kalicz and Makkay, 1977) — Szakálhát

*387. Szentes-Kistöke, co. Csongrád (Schupiter, 1931; Csalog, 1955; Kalicz and Makkay, 1977) — Szakálhát

*388. Szentes-Nagyhegy, co. Csongrád (Schupiter, 1931; Csallány, 1939; Kalicz and Makkay, 1977) — Szakálhát

*389. Szentes-Megyeháza, co. Csongrád (Schupiter, 1931; Kalicz and Makkay, 1977) — Szakálhát

390. Szentistván-Fehérlói csárda, co. Borsod
(Kalicz and Makkay, 1977)

*391. Szerencs-Hajdurét, co. Borsod — Middle, Late AVK, Tiszadob, Bükk
(Korek and Patay, 1958; Kalicz and Makkay, 1977)

*392. Szerencs-Hajdurét, co. Borsod — Szatmár II
(Kalicz and Makkay, 1972a; 1977)

393. Szerep-Balázsné halma, co. Hajdu-Bihar
(Mesterházy, 1970)

*394. Szilvásvárad-Istállóskői barlang, co. Heves — Late AVK, Tiszadob, Bükk
(Saád, 1929; Tompa, 1929; Korek, 1955;
Korek and Patay, 1958; Kalicz and Makkay, 1977)

*395. Szilvásvárad-Szalajka völgy, co. Heves — Late AVK, Bükk
(Salamon, 1960; 1963; 1970; Kalicz and Makkay,
1977)

*396. Szirmabesenyő-Homokbánya III, co. Borsod — Late AVK, Tiszadob
(Korek and Patay, 1958; Kemenczei and Végh,
1968; 1969; Kalicz and Makkay, 1977)

397. Szolnok-Tisza Antal TSZ, co. Szolnok
(Kalicz and Makkay, 1977)

398. Szolnok-Tiszapart közelében, co. Szolnok
(Márton, 1909; Kalicz and Makkay, 1977)

*399. Szolnok-Tüzköves, co. Szolnok — Szakálhát
(Stanczik, 1969; 1970; Kalicz and Makkay, 1977)

400. Szolnok-Villanytelep, co. Szolnok
(Csalog, 1955; Korek and Patay, 1956; Kalicz and
Makkay, 1977)

401. Tákos-Homokbánya, co. Szabolcs
(Kalicz and Makkay, 1977)

402. Tállya-Halastó-Szőlő, co. Borsod
(Gállus, 1936; Tompa, 1937; Korek and Patay,
1958; Kalicz and Makkay, 1977)

*403. Tápé-Lebő, co. Csongrád — Szakálhát
(Móra, 1930; Trogmayer, 1958; 1968; 1971;
Kalicz and Makkay, 1977)

404. Tapiószág-Zsiger, Co. Pest
(Tompa, 1937; Kalicz and Makkay, 1977)

*405. Tarcal-Allami Gazdaság Csemete kertje, co. Borsod — Early
(Kalicz and Makkay, 1977)

406. Tard, co. Borsod
(Kalicz and Makkay, 1977)

*407.	Tarnabod-Báb, co. Heves (Kalicz and Makkay, 1977)	Middle
*408.	Tarnabod-Csörszárok, co. Heves (Kalicz, 1969; Kalicz and Makkay, 1977)	Middle
*409.	Tarnabod-Kisberek-Kenderföldek, co. Heves (Kalicz, 1961; 1969; Kalicz and Makkay, 1977)	Middle
*410.	Tarnabod-Nagykert, co. Heves (Kalicz, 1969; Kalicz and Makkay, 1977)	Late AVK, Szakálhát
*411.	Tarnabod-Szilvás, co. Heves (Kalicz, 1969; Kalicz and Makkay, 1977)	Middle
*412.	Tarnabod-Templomföld, co. Heves (Kalicz, 1961; 1969; Kalicz and Makkay, 1977)	Middle, Late AVK, Szakálhát
*413.	Tarnabod-Ürgehát, co. Heves (Kalicz, 1969; Kalicz and Makkay, 1977)	Middle
*414.	Tarnaméra-Agyagbánya, co. Heves (Kalicz, 1969; Kalicz and Makkay, 1977)	Middle
*415.	Tarnaméra-Cselőházi homokbánya, co. Heves (Szabó, 1962; Kalicz, 1969; Kalicz and Makkay, 1977)	Late AVK
*416.	Tarnaméra-Tarna melletti homokbánya, co. Heves (Kalicz, 1969; Kalicz and Makkay, 1977)	Late AVK
*417.	Tarnaméra-Uszoda, co. Heves (Kalicz, 1969; Kalicz and Makkay, 1977)	Middle
418.	Tarnaméra-Zsákai Sándor gyümölcsöse, co. Heves (Kalicz, 1969; Kalicz and Makkay, 1977)	
*419.	Tarnaszadány-Sándorrésze, co. Heves (Kalicz, 1964; 1970; Kalicz and Makkay, 1977)	Middle, Late AVK, Szakálhát
*420.	Tépe-Kállo csatorna partja, co. Hajdu-Bihar (Kalicz and Makkay, 1977)	Middle
*421.	Tiszabábolna-Tanyai iskola, co. Borsod (Kalicz and Makkay, 1977)	Middle
*422.	Tiszabábolna-Tetves, co. Borsod (Kalicz and Makkay, 1977)	Middle
*423.	Tiszabercel- Keritőpart, co. Szabolcs (Korek, 1957; Korek and Patay, 1958; Kalicz and Makkay, 1977)	Tiszadob, Bükk I
*424.	Tiszabezdéd-Serpáva, co. Szabolcs (Kalicz and Makkay, 1972a; 1977)	Szatmár I
*425.	Tiszacsege-Homokbánya, co. Hajdu-Bihar (Kalicz and Makkay, 1972a; 1977)	Szatmár II

*426.	Tiszacsege-Hortobágy-Nagykecskés puszta, co. H-B (Éri and Kőszegi, 1957; Korek, 1960; Kalicz and Makkay, 1977)	Middle
427.	Tiszadada, co. Szabolcs (Tompa, 1929; Korek and Patay, 1958; Kalicz and Makkay, 1977)	
428.	Tiszadada-Kálvinháza, co. Szabolcs (Kalicz and Makkay, 1977)	
*429.	Tiszadob-Borzik-Zsidótábla, co. Szabolcs (Petróczy, 1945; Kalicz, 1956; Korek and Patay, 1958; Kalicz and Makkay, 1977)	Middle, Tiszadob, Bükk
*430.	Tiszadob-Ókenéz-Sziget I-II, co. Szabolcs (Petróczy, 1945; Csallány, 1958; Kalicz and Makkay, 1977)	Middle, Tiszadob
*431.	Tiszadorogma-Gépállomás, co. Borsod (Kalicz and Makkay, 1977)	Middle
432.	Tiszadorogma-Kövesút, co. Borsod (Kalicz and Makkay, 1977)	
433.	Tiszadorogma-Véggát, co. Borsod (Kalicz and Makkay, 1977)	
434.	Tiszaeszlár, co. Szabolcs (Korek and Patay, 1958; Kalicz and Makkay, 1977)	
435.	Tiszaföldvár, co. Szolnok (Kalicz and Makkay, 1977)	
436.	Tiszaföldvár-Sziget-Kőszegi legelő, co. Szolnok (Kalicz, 1957; Kalicz and Makkay, 1977)	
*437.	Tiszaföldvár-Téglagyár, co. Szolnok (Kalicz and Makkay, 1977)	Middle, Szakálhát
*438.	Tiszafüred-Ásotthalom, co. Szolnok (Kalicz and Makkay, 1972a; 1977)	Early
*439.	Tiszagyenda-Garahalom, co. Szolnok (Kalicz and Makkay, 1977)	Early
*440.	Tiszaigar-Csikos tanya, co. Szolnok (Korek, 1960; Kalicz and Makkay, 1977)	Szilmeg, Szakálhát
*441.	Tiszainoka-Szigetpart II, co. Szolnok (Kalicz, 1957; Kalicz and Makkay, 1977)	Late AVK
*442.	Tiszainoka-Szigetpart IV, co. Szolnok (Kalicz, 1957; Kalicz and Makkay, 1977)	Szakálhát
*443.	Tiszakarád-Bábahomoka, co. Borsod (Kalicz and Makkay, 1977)	Tiszadob

*444. Tiszakeszi-Szódadomb, co. Borsod (Tompa, 1937; Végh and Kemenczei, 1965; Kalicz and Makkay, 1977)	Middle
*445. Tiszakeszi-Tiszapart II, co. Borsod (Végh and Kemenczei, 1965; Kalicz and Makkay, 1977)	Middle
*446. Tiszakeszi-Tiszapart III, co. Borsod (Végh and Kemenczei, 1965; Kalicz and Makkay, 1977)	Early
*447. Tiszakeszi-Uj Elet TSZ, co. Borsod (Kalicz and Makkay, 1977)	Middle
*448. Tiszakürt-Műút III, co. Szolnok (Kalicz and Makkay, 1977)	Middle
*449. Tiszakürt-Tópart, co. Szolnok (Kalicz, 1957; Kalicz and Makkay, 1977)	Middle, Szakálhát
*450. Tiszaladány-Dohányos tanya, co. Borsod (Korek and Patay, 1958; Kalicz and Makkay, 1977)	Middle, Bükk
451. Tiszalök-Ciberéspart, co. Szabolcs (Korek and Patay, 1958; Kalicz and Makkay, 1977)	
452. Tiszalök-Kisfástanya-Bereczki dűlő, co. Szabolcs (Csallány and Gombás, 1965; Kalicz and Makkay, 1977)	
453. Tiszalök-Vásárhalom, co. Szabolcs (Kalicz and Makkay, 1977)	
*454. Tiszalúc-Vályogos, co. Borsod (Kalicz, 1961; Kalicz and Makkay, 1977)	Late AVK
455. Tiszanána, co. Heves (Kalicz and Makkay, 1977)	
456. Tiszaörvény, co. Szolnok (Tompa, 1937; Kalicz and Makkay, 1977)	
*457. Tiszaörvény-Temetődomb, co. Szolnok (Korek, 1960; Kalicz and Makkay, 1977)	Szatmár II
*458. Tiszaroff-Szakadópart-Alsórét, co. Szolnok (Korek and Patay, 1958; Kalicz and Makkay, 1977)	Middle, Szakálhát
*459. Tiszasas-Beltelek, co. Szolnok (Kalicz and Makkay, 1977)	Szakálhát
*460. Tiszasas-Rév I, co. Szolnok (Kalicz, 1957; Kalicz and Makkay, 1977)	Late AVK, Szakálhát
461. Tiszasziget-Kissziget I-III, co. Csongrád (Banner, 1928; Banner and Párducz, 1948; Kalicz and Makkay, 1977)	

462. Tiszaszőlős-Csákányszeg, co. Szolnok (Csalog, 1961; Kalicz and Makkay, 1977)	
463. Tiszaszőlős-Csákányszeg-10. lelőhely, co. Szolnok (Kalicz and Makkay, 1977)	
464. Tiszaszőlős-Csontospart III, co. Szolnok (Csalog, 1964; Kalicz and Makkay, 1972a; 1977)	
465. Tiszaszőlős-Telekpart, co. Szolnok (Kalicz and Makkay, 1977)	
466. Tiszatelek-Csókalapos, co. Szabolcs (Kalicz and Makkay, 1977)	
467. Tiszatelek-Csókatag, co. Szabolcs (Kalicz and Makkay, 1977)	
*468. Tiszatenyő-Sziget-Bátor tanya, co. Szolnok (Kalicz and Makkay, 1977)	Szakálhát
*469. Tiszaug-Tópart II, co. Szolnok (Szabó, 1934; Kalicz, 1957; Kalicz and Makkay, 1977)	Szakálhát
*470. Tiszavalk-Négyesi határ, co. Borsod (Korek, 1970; Patay, 1969; Kalicz and Makkay, 1972a; 1977)	Szatmár II
*471. Tiszavalk-Tetes, co. Borsod (Kalicz and Makkay, 1977)	Late AVK
*472. Tiszavárkony-Hugyinpart, co. Szolnok (Korek, 1960; Kalicz and Makkay, 1977)	Middle, Szakálhát
*473. Tiszavárkony-Szőlő, co. Szolnok (Kalicz and Makkay, 1977)	Szakálhát
*474. Tiszavasvári-Badicshát, co. Szabolcs (Kalicz and Makkay, 1977)	Middle
*475. Tiszavasvári-Bereczki halom, co. Szabolcs (Csalog, 1955; 1956; Kalicz, 1956; Kalicz and Makkay, 1977)	Late AVK, Szilmeg, Bükk
*476. Tiszavasvári-Fejérszik, co. Szabolcs (Kalicz and Makkay, 1977)	Middle, Bükk
477. Tiszavasvári-Iskola dűlő, co. Szabolcs (Kalicz and Makkay, 1977)	
*478. Tiszavasvári-Józsefháza-Téglás, co. Szabolcs (Kalicz and Makkay, 1977)	Tiszadob
*479. Tiszavasvári-Keresztesi partok, co. Szabolcs (Kalicz, 1956; Kalicz and Makkay, 1977)	Tiszadob
*480. Tiszavasvári-Keresztfal, co. Szabolcs (Csallány and Gombás, 1963; 1964; 1965; Kalicz and Makkay, 1966; 1977; Kalicz, 1970)	Late AVK, Tiszadob, Esztár, Bükk

481. Tiszavasvári-Kisasszonydülö, co. Szabolcs (Csallány and Gombás, 1963; Kalicz and Makkay, 1977)	
*482. Tiszavasvári-Paptelekhát, co. Szabolcs (Korek, 1957; Kalicz, 1956; 1958; Kutzián, 1963; Kalicz and Makkay, 1977)	Middle, Late AVK, Tiszadob, Bükk I
*483. Tiszavasvári-Utaséri dülö, co. Szabolcs (Kalicz and Makkay, 1977)	Bükk II
*484. Tolcsva-Bányadülö, co. Borsod (Kalicz and Makkay, 1977)	Late AVK, Tiszadob, Bükk III
*485. Tolcsva-Előhegy, co. Borsod (Kalicz and Makkay, 1977)	Late AVK, Tiszadob, Bükk I
*486. Tolcsva-Pénzásó dülö, co. Borsod (Kalicz, 1959; Kalicz and Makkay, 1977)	Tiszadob
*487. Tolcsva-Rány, co. Borsod (Kalicz and Makkay, 1977)	Bükk
*488. Tolcsva-Várhegy, co. Borsod (Kalicz, 1958; Kalicz and Makkay, 1977)	Bükk III
*489. Törökszentmiklós-Surján, co. Szolnok (Kalicz and Makkay, 1977)	Middle
*490. Tószeg-Laposhalom, co. Szolnok (Banner, Bóna and Márton, 1959; Kalicz and Makkay, 1977)	Middle, Tiszadob
*491. Tószeg-Telek, co. Szolnok (Tompa, 1929; Banner, Bóna and Márton, 1959; Kalicz and Makkay, 1977)	Szakálhát
492. Tótkómlós-Viharsarok TSZ, co. Békés (Kalicz, 1967; Kalicz and Makkay, 1977)	
493. Túrkeve-Himesaljai lapos, co. Szolnok (Korek, 1960; Kalicz and Makkay, 1977)	
*494. Túrkeve-Lyukashalom, co. Szolnok (Korek, 1960; Kalicz and Makkay, 1977)	
*495. Túrkeve-Turkeddi halom, co. Szolnok (Korek 1960; Kalicz and Makkay, 1977)	Middle
*496. Tuzsér-Belterület, co. Szabolcs (Kalicz and Makkay, 1977)	
*497. Ujfehértó, co. Szabolcs (Kalicz and Makkay, 1977)	Esztár
498. Ujszentmargita, co. Hajdu-Bihar (Kalicz and Makkay, 1977)	

*499. Ujtikos-Demeterkút-Agyagbánya, co. Hajdu-Bihar (Kutzián, 1946; Korek, 1960; Kalicz and Makkay, 1977)	Middle, Tiszadob
500. Ujtikos-Kisfaludi erdő , co. Hajdu-Bihar (Kalicz and Makkay, 1977)	
*501. Ujtikos-Kuhn tanya, co. Hajdu-Bihar (Kalicz and Makkay, 1977)	Middle
*502. Ujtikos-Néptelep, co. Hajdu-Bihar (Korek, 1960; Kalicz and Makkay, 1977)	Middle
503. Ujtikos-Rakodópart, co. Hajdu-Bihar (Korek, 1960; Kalicz and Makkay, 1977)	
*504. Ujtikos-Tikos domb, co. Hajdu-Bihar (Korek, 1959; 1960; Kalicz and Makkay, 1977)	Middle, Tiszadob, Esztár, Bükk I
505. Uppony-1. sz. kőfülke, co. Borsod (Korek and Patay, 1958; Kalicz and Makkay, 1977)	
*506. Uppony-Malomgát, co. Borsod (Korek, 1962; 1970; Kalicz and Makkay, 1977)	Middle
*507. Uppony-Mogyorós oldal, co. Borsod (Korek, 1962; 1970; Kalicz and Makkay, 1977)	Middle
*508. Vadna-Agyagbánya, co. Borsod (Korek, 1957; 1960; Korek and Patay, 1958; Kalicz and Makkay, 1977)	Late AVK, Bükk
*509. Váncsod-Berettyópart, co. Hajdu-Bihar (Kalicz and Makkay, 1977)	Esztár
*510. Varbó-Lambrecht-Kálmán barlang, co. Borsod (Vértes, 1953; Korek and Patay, 1958; Kalicz and Makkay, 1977)	Bükk
*511. Vásárosnamény-Farönk telep, co. Szabolcs (Korek, 1964; Kalicz and Makkay, 1977)	Esztár
512. Vasmegyer-Kisvármi-Nagyvármi, co. Szabolcs (Korek and Patay, 1958; Kalicz and Makkay, 1977)	
513. Vasmegyer-Megyalja, co. Szabolcs (Kalicz and Makkay, 1977)	
*514. Vencsellő-Lónyai csatorna, co. Szabolcs (Kalicz, 1956; 1957a; Kalicz and Makkay, 1977)	Middle
515. Vésztő-Hidashát-Hosszúhalom , co. Békés (Kalicz and Makkay, 1977)	
*516. Vésztő-Mágorhalom, co. Békés (Hegedüs, pers. comm.)	Szakálhát

*517.	Viss-Csenkeoldal, co. Szabolcs (Korek and Patay, 1958; Kalicz and Makkay, 1977)	Late AVK, Tiszadob, Bükk I
*518.	Visznek-Kecskedomb, co. Heves (Kalicz and Makkay, 1977)	Szakálhát
*519.	Zagyvarékás-Jánosi Imre tanya, co. Szolnok (Kalicz and Makkay, 1977)	Szakálhát
520.	Zagyvarékás-Páskum I, co. Szolnok (Kalicz and Makkay, 1977)	
521.	Zagyvarékás-Páskum II, co. Szolnok (Kalicz and Makkay, 1977)	
*522.	Zagyvarékás-Páskum-Szélmalom, co. Szolnok (Kalicz and Makkay, 1977)	Szakálhát
*523.	Zalkod-Csenkeoldal, co. Borsod (Salamon, 1966; Kalicz, 1970; Kalicz and Makkay, 1977)	Szilmeg, Bükk III
*524.	Zalkod-Jakabdomb, co. Borsod (Kalicz, 1970; Kalicz and Makkay, 1977)	Middle, Tiszadob, Bükk I
*525.	Zaránk, co. Heves (Kalicz and Makkay, 1977)	Middle
*526.	Zsáka-Markó, co. Hajdu-Bihar (Kalicz and Makkay, 1977)	Late AVK, Esztár
*527.	Zsáka-Vekerdi út-Gombás ér, co. Hajdu-Bihar (Mesterházy, 1966; Kalicz and Makkay, 1977)	Middle, Szilmeg
*528.	Zsáka-Vizesi tanya, co. Hajdu-Bihar (Sőregi, 1935; Kutzián, 1963; Kalicz and Makkay, 1977)	Late AVK, Esztár

Transdanubian Linear Pottery culture

		Chronology
* 1.	Aba-Felsőszentiván-Ángyihegy, co. Fejér (Petres, 1962; Fitz, 1963-64; Bánki, 1967-68; Makkay, 1970)	Early
* 2.	Adorjánháza-no. 3, co. Veszprém (Bakay et al., 1970)	Middle
* 3.	Adorjánháza-no. 5, co. Veszprém (Bakay et al., 1970)	Middle
* 4.	Adorjánháza-no. 6, co. Veszprém (Bakay et al., 1970)	Middle
* 5.	Ajka-no. 2, co. Veszprém (Bakay et al., 1970)	Middle
* 6.	Almásfüzitő-Foktorok, co. Komárom (Vadasz, 1972)	Zseliz
* 7.	Alsópáhok-no. 1, co. Veszprém (Bakay et al., 1966)	Middle
* 8.	Alsópáhok-no. 6, co. Veszprém (Bakay et al., 1966)	Middle
* 9.	Alsópáhok-Káthánalja, co. Veszprém (Bakay et al., 1966)	Middle
*10.	Andocs-Nagytoldi puszta, co. Somogy (Draveczky, 1966; 1970)	Middle
11.	Andocs-Ujhalastó, co. Somogy (Draveczky, 1966; Draveczky and Garam, 1967)	
*12.	Aszód, co. Pest (Kalicz, 1967a)	Notenkopf, Zseliz
*13.	Aszód-Tófalu, co. Pest (Kalicz, pers. comm.)	Notenkopf, Zseliz
*14.	Bakonygyirót-Felsőgyöri föld, co. Veszprém (Dax et al., 1972)	Middle
*15.	Bakonyság-Öregirtás, co. Veszprém (Dax et al., 1972)	Middle
*16.	Balatonakali-Ság puszta-Téglagyár, co. Veszprém (Éri et al., 1969)	Zseliz

*17.	Balatonakali-Sósi földek, co. Veszprém (Éri et al., 1966)	Middle
*18.	Balatonederics-no. 3, co. Veszprém (Bakay et al., 1966)	Middle
*19.	Balatonendréd-Vakláp,u., co. Somogy (Kocztur, 1964; Draveczky, 1966; 1970; Draveczky and Sági, 1964)	Middle
*20.	Balatonfüred-Siskei templomrom, co. Veszprém (Éri et al., 1969)	Middle
*21.	Balatongyörök-Becehelyi major, co. Veszprém (Bakay et al., 1966)	Middle
*22.	Balatonkenese-Akarattya II, co. Veszprém (Éri et al., 1969)	Middle
*23.	Balatonkenese-Sándorhegy, co. Veszprém (Éri et al., 1969)	Middle
*24.	Balatonszabadi-Sáfránykert, co. Somogy (Kocztur, 1964; Dravetzky, 1966; 1970)	Middle
*25.	Balatonszepezd-no. 8, co. Veszprém (Bakay et al., 1966)	Middle
*26.	Barnag-Remete kert, co. Veszprém (Éri et al., 1969)	Middle, Zseliz
27.	Becsehely, co. Zala (Tompa, 1937; Quitta, 1960; Korek, 1960)	
*28.	Becsvölgye-Barabásszeg, co. Zala (Szentmihályi, 1954)	Middle
*29.	Bedegkér-Tótkértöl délre, co. Somogy (Kocztur, 1964; Draveczky, 1970)	Middle
30.	Bélavár, co. Somogy (Bándi, pers. comm.)	
*31.	Bicske-Galagonyáshegy, co. Fejér (Dornyay and Csalog, 1945; Petres, 1954; 1959; Csalog, 1955; Quitta, 1960; Makkay, 1966; 1970; 1972; pers. comm.)	Early, Notenkopf
*32.	Bodajk, co. Fejér (Petres, 1954; Kralovánszky, 1961-62; Kiss, 1961-62)	Middle
*33.	Bonyhád, co. Tolna (Banner, 1943; Csalog, 1942; Quitta, 1960)	Middle
*34.	Budakalász, co. Pest (Kalicz, pers. comm.)	Notenkopf

*35.	Budapest-Békásmegyer-Vöröshadsereg u., co. Pest (Tompa, 1937; Petres, 1954; Csalog, 1955; Quitta, 1960; Csánk, 1964; 1967)	Zseliz
*36.	Budapest-Kossuth L. Üdülőpart, co. Pest (Nagy, 1960)	Zseliz
*37.	Budapest-XXII. Nagytétény, co. Pest (Gallus, 1936; Tompa, 1937; Petres, 1954; Korek and Patay, 1958)	Zseliz
38.	Budapest-Óbuda, co. Pest (Tompa, 1942)	
*39.	Budapest-III. Pünkösd fürdő, co. Pest (Nagy, 1960)	Zseliz
*40.	Budapest-Tabán, co. Pest (Tompa, 1937; Quitta, 1960)	Zseliz
*41.	Büssü, co. Somogy (Bándi, pers. comm.)	Middle
*42.	Cserszegtomaj-Bikádtető, co. Veszprém (Bakay et al., 1966)	Middle
43.	Csibrák, co. Tolna (Bándi, pers. comm.)	
*44.	Csikvánd-Csillagtag, co. Veszprém (Dax et al., 1972)	Middle
*45.	Csikvánd-Kalmári tag, co. Veszprém (Dax et al., 1972)	Middle
*46.	Csögle-no. 3, co. Veszprém (Bakay et al., 1970)	Middle
*47.	Dabronc-no. 2, co. Veszprém (Bakay et al., 1970)	Middle
*48.	Dabronc-Rétalja dülö, co. Veszprém (Bakay et al., 1970)	Middle
*49.	Dabronc-no. 6, co. Veszprém (Bakay et al., 1970)	Middle
50.	Döbrököz, co. Tolna (Bándi, pers. comm.)	
*51.	Dunaújváros-Papsziget, co. Fejér (Makkay, 1970)	Zseliz
*52.	Dunaújváros-Rácdomb, co. Fejér (Makkay, 1970)	Zseliz
53.	Dúzs, co. Tolna (Bándi, pers. comm.)	

54.	Ecsény, co. Somogy (Draveczky and Sági, 1964; Draveczky, 1966)	
*55.	Egyházaskeszö-Molnársziget, co. Veszprém (Dax et al., 1972)	Middle, Zseliz
*56.	Egyházaskeszö-Kurcant, co. Veszprém (Dax et al., 1972)	Middle
57.	Egyházaskozár, co. Baranya (Bándi, pers. comm.)	
58.	Értény, co. Tolna (Bándi, pers. comm.)	
59.	Esztergom, co. Komárom (Kalicz and Makkay, 1972b)	
*60.	Esztergom-Helemba sziget, co. Komárom (Trogmayer, 1960)	Zseliz
*61.	Etyek-Kálvária domb, co. Fejér (Bánki, 1966; 1967-68)	Zseliz
*62.	Fajsz-Garadomb, co. Bács-Kiskun (Horváth, 1972)	
*63.	Fehérvárcsurgó-Igarpuszta, co. Fejér (Makkay, 1968; Bánki, 1969)	Zseliz
*64.	Fehérvárcsurgó-Igarpuszta-Homokbánya, co. F. (Makkay, 1968)	Zseliz
65.	Felsönyék, co. Tolna (Bándi, pers. comm.)	
*66.	Felsöörs-Ányosháza, co. Veszprém (Éri et al., 1969)	Middle
*67.	Felsöörs-TSZ major, co. Veszprém (Éri et al., 1969)	Middle
*68.	Fonyód-Bézsenypuszta-Tözegtelep, co. Somogy (Kocztur, 1964; Draveczky, 1964; 1970)	Middle, Zseliz
*69.	Galgahéviz-Monostori dülö, co. Pest (Kalicz, 1967a)	Notenkopf, Zseliz
*70.	Gógánfa-Nyires, co. Veszprém (Bakay et al., 1970)	Middle
*71.	Gógánfa-no. 6, co. Veszprém (Bakay et al., 1970)	Zseliz
*72.	Gógánfa-no. 9, co. Veszprém (Bakay et al., 1970)	Middle
*73.	Gógánfa-no. 10, co. Veszprém (Bakay et al., 1970)	Middle

*74.	Góganfa-no. 11, co. Veszprém (Bakay et al., 1970)	Middle
*75.	Gölle, co. Somogy (Pusztai, 1958; Kocztur, 1964; Draveczky and Sági, 1964)	Middle
*76.	Görgeteg-Szőlőhegy, co. Somogy (Korek, 1963; Kocztur, 1964; Draveczky, 1966; 1970)	Middle
*77.	Gyenesdiás-Lasics F. telke, co. Veszprém (Bakay et al., 1966)	Middle
*78.	Győr-Pándzsa dülő-Pápai vám, co. Győr-Sopron (Mithay, 1954; 1956; 1966)	Middle, Zseliz
79.	Gyügy, co. Somogy (Bándi, pers. comm.)	
*80.	Gyulafirátót-TSZ major, co. Veszprém (Éri et al., 1969)	Middle
81.	Gyulaj, co. Tolna (Bándi, pers. comm.)	
*82.	Harc-Nyanyapuszta, co. Tolna (Wosinszky, 1896; Kalicz and Makkay, 1972a; 1972b)	Medina
*83.	Hatvan, co. Pest (Kalicz, pers. comm.)	Notenkopf, Zseliz
*84.	Hévizgyörk-Kaparóházi dülő, co. Pest (Kalicz, 1966; 1970)	Notenkopf
*85.	Hidegkút-Vörösföldek, co. Veszprém (Éri et al., 1969)	Middle
*86.	Hidegkút-Középdülő, co. Veszprém (Éri et al., 1969)	Middle
*87.	Hidegkút-Lincakert, co. Veszprém (Éri et al., 1969)	Middle
88.	Högyész, co. Tolna (Bándi, pers. comm.)	
89.	Iregszemcse, co. Tolna (Bándi, pers. comm.)	
*90.	Iszkáz-no. 2, co. Veszprém (Bakay, et al., 1970)	Zseliz
*91.	Iszkáz-no. 3, co. Veszprém (Bakay et al., 1970)	Middle
*92.	Iszkáz-no. 4, co. Veszprém (Bakay et al., 1970)	Zseliz

*93.	Iszkáz-no. 6, co. Veszprém (Bakay et al., 1970)	Zseliz
*94.	Juta-Gagarin dülö, co. Somogy (Gazdapusztai and Draveczky, 1967; Draveczky, 1970)	Middle
*95.	Kádárta-Lánci dülö, co. Veszprém (Éri et al., 1969)	Middle
96.	Kajdacs, co. Tolna (Bándi, pers. comm.)	
*97.	Kamond-no. 7, co. Veszprém (Bakay et al., 1970)	Middle
*98.	Kamond-no. 10, co. Veszprém (Bakay et al., 1970)	Middle
99.	Kánya, co. Tolna (Bándi, pers. comm.)	
100.	Kápolnásnyék, co. Fejér (Bándi, pers. comm.)	
101.	Kaposvár-Ólaki dülö, co. Somogy (Csalog, 1955; Quitta, 1960; Korek, 1963; Kocztur, 1964; Draveczky, 1966; 1970; Draveczky and Ságı, 1964)	
*102.	Kaposvár-Téglagyár-Arany J.u., co. Somogy (Draveczky, 1966)	Middle
*103.	Kaposvár-Tüskevár-Vasutállomás, co. Somogy (Draveczky, 1966)	Middle
*104.	Karád-Berencse, co. Somogy (Draveczky, 1966; 1970)	Middle
*105.	Karmacs-no. 6, co. Veszprém (Bakay et al., 1966)	Middle
106.	Kartal, co. Pest (Kalicz, 1967)	
*107.	Kerta-no. 4, co. Veszprém (Bakay et al., 1970)	Middle
*108.	Keszthely-no. 6, co. Veszprém (Bakay et al., 1966)	Middle
*109.	Keszthely-Fenékpuszta, co. Veszprém (Bakay et al., 1966)	Middle
*110.	Keszthely-no. 32, co. Veszprém (Bakay et al., 1966)	Middle
*111.	Keszthely-Balatoni Múzeum, co. Veszprém (Bakay et al., 1966)	Middle

*112.	Keszthely-Méntelep, co. Veszprém (Bakay et al., 1966)	Middle
*113.	Keszthely-Várkert, co. Veszprém (Bakay et al., 1966)	Middle
*114.	Keszthely-Nagy L.u. 54, co. Veszprém (Dornyay and Csalog, 1945; Csalog, 1955; Bakay et al., 1966)	Middle
*115.	Keszthely-Ujdülö, co. Veszprém (Bakay et al., 1966)	Middle
*116.	Keszthely-Zsidi út, co. Veszprém (Horváth, 1970)	Zseliz
*117.	Kéthely-Baglyashegy, co. Somogy (Kocztur, 1964; Draveczky, 1966; 1970)	Middle
*118.	Kiscsősz-no. 7, co. Veszprém (Bakay et al., 1970)	Middle
119.	Kiskánya, co. Tolna (Csalog, 1941b; Quitta, 1960; Kalicz and Makkay, 1972b)	
120.	Köröshegy, co. Somogy (Bándi, pers. comm.)	
121.	Kötcse, co. Somogy (Bándi, pers. comm.)	
*122.	Külsövat-Derma III, co. Veszprém (Dax et al., 1972)	Middle
*123.	Külsövat-Gányi tó, co. Veszprém (Dax et al., 1972)	Middle, Zseliz
*124.	Külsövat-Bánhalmi Ág. gyümölcsöse, co. V. (Dax et al., 1972)	Middle
*125.	Külsövat-Belsöbánd, co. Veszprém (Dax et al., 1972)	Middle
*126.	Kunszentmiklós-Középtamás, co. Bács-Kiskun (Tóth, 1967)	
*127.	Kustánszeg, co. Zala (Kalicz, pers. comm.)	Middle
*128.	Lengyeltóti-Nagytatársánc, co. Somogy (Kocztur, 1964; Draveczky, 1966; 1970)	Middle, Zseliz
*129.	Lesencefalu-no. 2, co. Veszprém (Bakay et al., 1966)	Middle
*130.	Letkés-Keresztföldek-Vizafogó, co. Pest (Papp, 1966; 1967; 1970)	Zseliz

*131. Litér-Kéktói dülö, co. Veszprém (Éri et al., 1969) — Middle

*132. Lovas-Királykút I, co. Veszprém (Éri et al., 1969) — Middle

*133. Lovasberény-Alsótelek-Gubadomb, co. Fejér (Petres, 1954; Bánki, 1969; Makkay, 1970) — Notenkopf

134. Mágocs, co. Baranya (Tompa, 1937; Quitta, 1960)

135. Magyarkeszi, co. Tolna (Bándi, pers. comm.)

*136. Marcaltö-Gerencei dülö, co. Veszprém (Dax et al., 1972) — Middle

*137. Medina-Siópart, co. Tolna (Banner, 1943; Tompa, 1937; Csalog, 1941b; Kalicz and Makkay, 1972) — Medina

*138. Megyer-Csombor, co. Veszprém (Bakay, et al., 1970) — Middle

*139. Mencshely-Murvagödrök, co. Veszprém (Éri et al., 1969) — Middle

*140. Mende, co. Pest (Patay, 1966-67) — Notenkopf, Zseliz

*141. Mezökomárom-Alsóhegy, co. Fejér (Makkay, 1970) — Middle

*142. Mihályfa-no. 8, co. Veszprém (Bakay et al., 1970) — Middle

*143. Monoszló-Boncosdülö, co. Veszprém (Bakay et al., 1966) — Middle

*144. Mór-Vajal dülö, co. Fejér (Makkay, 1970) — Middle

145. Mosonszentjános-Hansági tanyák, co. Györ-So. (Pusztai, 1962)

*146. Mosonszentmiklos-Akasztódomb, co. Györ-So. (Uzsoki and Gábler, 1967) — Middle

*147. Mözs, co. Tolna (Salamon, 1962; 1970) — Middle

*148. Nagyalásony-no. 5, co. Veszprém (Bakay et al., 1970) — Middle

*149. Nagyberki-Szalacska, co. Somogy (Dornyay and Csalog, 1945; Kocztur, 1964; Draveczky, 1966; 1970) — Middle

*150.	Nagyhörcsök-Törökdomb, co. Fejér (Bánki, 1969; 1970)	Middle
151.	Nagykónyi, co. Tolna (Bándi, pers. comm.)	
152.	Nagyszakoly, co. Tolna (Bándi, pers. comm.)	
*153.	Nagytarcsa, co. Pest (Bakay, 1970)	
*154.	Nagyvászony-Baráti dülö, Csapás I, co. Ves. (Éri et al., 1969)	Middle
*155.	Nagyvázsony-Szérüskerti dülö, co. Veszprém (Éri, 1962; Éri et al., 1969)	Middle, Zseliz
*156.	Nemesgörzsöny-Szélmezöi dülö, co. Ves. (Dax et al., 1972)	Middle
*157.	Nemesszalók-Rövid dülö, co. Veszprém (Dax et al., 1972)	Middle
*158.	Neszmély-Tekeres patak, co. Komárom (Makkay, 1960; Kutzián, 1963)	Zseliz
*159.	Óhid-no. 8, co. Veszprém (Bakay et al., 1970)	Middle
*160.	Örvényes-Kukorica föld, co. Veszprém (Éri et al., 1969)	Middle, Zseliz
*161.	Ösi-Buhinvölgy II, co. Veszprém (Éri et al., 1969)	Middle
*162.	Ösi-Zsigmondpuszta I, co. Veszprém (Éri et al., 1969)	Middle
*163.	Osztopán-Balicskai dülö, co. Somogy (Dornyay and Csalog, 1945; Kocztur, 1964; Draveczky, 1966; 1970)	Middle
164.	Pamuk, co. Somogy (Kocztur, 1964; Draveczky, 1966)	
*165.	Papkeszi-Malom, co. Veszprém (Kiss, 1957; Quitta, 1960; Éri et al., 1969)	Middle
*166.	Pári-Altacker, co. Tolna (Torma, 1969; 1971)	Middle, Notenkopf, Zseliz
167.	Pécel, co. Pest (Korek, 1960)	
168.	Pécs-Középhegy dülö, co. Baranya (Bándi, pers. comm.)	

*169.	Pécsely-Zádorvár, co. Veszprém (Éri et al., 1969)	Middle, Zseliz
*170.	Pécsely-Bánya, co. Veszprém (Dax, 1967; Éri et al., 1969)	Middle, Zseliz
171.	Pécsmegyer-Rét, co. Baranya (Draveczky, 1966)	
*172.	Pécsvárad-Aranyhegy, co. Baranya (Tompa, 1937; Dombay, 1958)	Middle
*173.	Penc-Sügyi puszta, co. Nógrád (Patay, 1956)	Zseliz
*174.	Pilismarót-Szobi rév, co. Komárom (Kralovánszky and Kovalovszki, 1960)	Zseliz
*175.	Pomáz-Posztógyár, co. Pest (Kalicz, pers. comm.)	Notenkopf, Zseliz
*176.	Pomáz-Zdravlyák dülö., co. Pest (Kutzián, 1958a)	Notenkopf, Zseliz
177.	Ráksi-Futball pálya, co. Somogy (Draveczky, 1966; 1970)	
*178.	Ráksi-Kertekalja, co. Somogy (Draveczky, 1966; 1970)	Middle
179.	Regöly-Füzfás, co. Somogy (Torma, 1968; 1970)	
*180.	Révfülöp-no. 1, co. Veszprém (Bakay et al., 1966)	Early
*181.	Rigács-no. 2, co. Veszprém (Bakay et al., 1970)	Middle
*182.	Ságvár-Jabavölgyi Ujmajor, co. Somogy (Kocztur, 1964; Draveczky, 1970)	Middle
*183.	Sárkeresztes-Pékmalom domb, co. Fejér (Makkay, 1969; 1970; Bánki, 1970)	Middle, Notenkopf
*184.	Sármellék-Égenföld, co. Veszprém (Bakay et al., 1966)	Middle
*185.	Sármellék-Belterület, co. Veszprém (Bakay et al., 1966)	Middle
*186.	Sármellék-no. 5, co. Veszprém (Bakay et al., 1966)	Zseliz
*187.	Sármellék-Lajosházi major, co. Veszprém (Bakay et al., 1966)	Middle
*188.	Sármellék-Devecseri dülö, co. Veszprém (Bakay et al., 1966)	Middle

*189.	Sármellék-no. 12, co. Veszprém (Bakay et al., 1966)	Middle
*190.	Sármellék-no. 14, co. Veszprém (Pekáry, 1954; Korek, 1968; Quitta, 1960; Bakay et al., 1966)	Middle
191.	Simontornya, co. Tolna (Csalogovits, 1930; Wosinsky, 1896)	
*192.	Sóly-Rétmelléki dülö, co. Veszprém (Éri et al., 1969)	Middle
193.	Somberek, co. Baranya (Bándi, pers. comm.)	
*194.	Somlószölös-no. 7, co. Veszprém (Bakay et al., 1970)	Middle
*195.	Somlóvásárhely-no. 18, co. Veszprém (Bakay et al., 1970)	Middle
196.	Somogyapáti, co. Baranya (Draveczky, 1966)	
*197.	Somogysámson, co. Somogy (Draveczky, 1966)	Middle
*198.	Somogyszil-Dögkúti dülö, co. Somogy (Draveczky, 1970; Draveczky and Sági, 1964)	Middle
199.	Somogyvámos-Templomrom, co. Somogy (Valter, 1969)	
200.	Somogyvár, co. Somogy (Bándi, pers. comm.)	
*201.	Sukoró-Tóradülö, co. Fejér (Makkay, 1969; 1970; Bánki, 1970)	Notenkopf, Zseliz
*202.	Süttö-Vasuti örház, co. Komárom (Kalicz, 1960)	Zseliz
203.	Szak, co. Komárom (Kiss, 1958)	
*204.	Szakály-TSZ major, co. Tolna (Torma, 1970a; 1971a)	Middle
205.	Szakcs, co. Tolna (Bándi, pers. comm.)	
*206.	Székesfehérvár--Szárazrét, co. Tolna (Bánki, 1970)	Notenkopf
*207.	Székesfehérvár-Székesegyház-Arany J.u., co. F. (Bánki, 1970; Makkay, 1970)	Early
208.	Szekszárd, co. Tolna (Bándi, pers. comm.)	

*209.	Szentendrei sziget, co. Pest (Kalicz, pers. comm.)	Notenkopf, Zseliz
*210.	Szentendrei sziget déli csúcsa, co. Pest (Kalicz, 1955)	Notenkopf, Zseliz
*211.	Szentgyörgyvár-Betefa, co. Veszprém (Bakay et al., 1966)	Middle
*212.	Szentimrefalva-no. 4, co. Veszprém (Bakay et al., 1970)	Middle
*213.	Szentjakabfa-Lázföld, co. Veszprém (Bakay et al., 1966)	Zseliz
*214.	Szentkirályszabadja-Romkút, co. Veszprém (Éri et al., 1969)	Middle, Zseliz
*215.	Szentlörinc-Téglagyár, co. Baranya (Bándi, 1969)	Middle
*216.	Szentmiklós-Szentmiklósi tagosztály, co. Pest (Kalicz, pers. comm.)	Zseliz
*217.	Szob-Öregfalu dülö, co. Pest (Petres, 1954)	Zseliz
218.	Szöny, co. Komárom (Csalog, 1955)	
*219.	Taliándörögd-Klastromdülö, co. Veszprém (Bakay et al., 1966)	Middle
*220.	Tamási-Vagóhid, co. Tolna (Torma, 1968; 1970)	Middle
*221.	Tapióbicske, co. Pest (Kalicz, pers. comm.)	Zseliz
*222.	Tatabánya-Kertváros, co. Komárom (Biró, 1961)	Zseliz
*223.	Tengelic-Felsötengelic puszta, co. Tolna (Kocztur, 1970)	Middle
224.	Tolnanémedi, co. Tolna (Bándi, pers. comm.)	
225.	Toponár, co. Somogy (Bándi, pers. comm.)	
*226.	Tótvázsony-Viztekerület, co. Veszprém (Éri, et al., 1969)	Middle
227.	Turbék, co. Baranya (Bándi, pers. comm.)	
*228.	Tüskevár-Sásréti dülö, co. Veszprém (Kiss, 1967)	Middle

*229.	Tűskevár-no. 10, co. Veszprém (Bakay et al., 1970)	Middle
230.	Újireg, co. Tolna (Bándi, pers. comm.)	
231.	Vác-Dunai Cement művek, co. Pest (Kőszegi, 1961)	
*232.	Várpalota-Hősi emlékmű, co. Veszprém (Éri et al., 1969)	Middle
*233.	Várpalota-Inota-Templomdűlő, co. Veszprém (Éri et al., 1969)	Middle
*234.	Várpalota-Hangyálos, co. Veszprém (Éri et al., 1969)	Middle, Zseliz
*235.	Vaszar-Tórét I, co. Veszprém (Dax et al., 1972)	Middle
*236.	Vászoly-Keretkalja-Kisdűlő, Buka, co. Vesz. (Éri et al., 1969)	Middle
*237.	Vászoly-Temető, co. Veszprém (Éri et al., 1969)	Middle
*238.	Veszprém-Várhegy, co. Veszprém (Kelemen and Tóth, 1966; Éri et al., 1969)	Middle, Zseliz
*239.	Veszprém-Tiszti lakások, co. Veszprém (Éri et al., 1969)	Zseliz
*240.	Veszprém-Hóvirág lakótelep, co. Veszprém (Csalogovits, 1930; Dornyay and Csalog, 1945; Éri et al., 1969)	Middle, Notenkopf
*241.	Veszprém-József A.u. 20, co. Veszprém (Szentleleky, 1960; Éri et al., 1969)	Middle, Zseliz
*242.	Veszprém-Nyúlkertek, co. Veszprém (Éri et al., 1969)	Middle
*243.	Veszprém-Sashegy I, co. Veszprém (Éri et al., 1969)	Middle
*244.	Veszprém-Somos II, co. Veszprém (Éri et al., 1969)	Middle
*245.	Veszprém-Juta-Diósdűlő, co. Veszprém (Éri et al., 1969)	Middle
*246.	Veszprémfajsz-Ányoskút, co. Veszprém (Éri et al., 1969)	Middle
*247.	Vilonya-Templomföldе, co. Veszprém (Éri et al., 1969)	Middle
*248.	Vonyarc Vashegy-Pintér villa, co. Veszprém (Bakay et al., 1966; alicz, pers. comm.)	Early

*249.	Vörösto-Ragonya, co. Veszprém (Éri et al., 1969)	Middle, Notenkopf, Zseliz
*250.	Vörs, co. Somogy (Kocztur, 1964; Draveczky, 1966; 1970)	Middle
*251.	Zalagyömörö-no. 1, co. Veszprém (Bakay et al., 1970)	Middle
*252.	Zalavár-Keleti J. udvara, co. Veszprém (Bakay et al., 1966)	Middle
*253.	Zalavár-Lantos Hilár háza, co. Veszprém (Bakay et al., 1966)	Middle
*254.	Zalavár-Hosszusziget, co. Veszprém (Bakay et al., 1966)	Middle
*255.	Zalavár-no. 36, co. Veszprém (Bakay et al., 1966)	Middle
*256.	Zalavár-no. 38, co. Veszprém (Bakay et al., 1966)	Middle
*257.	Zánka-no. 7, co. Veszprém (Bakay et al., 1966)	Middle
*258.	Zánka-no. 10, co. Veszprém (Bakay et al., 1966)	Middle
*259.	Zengövárkony, co. Baranya (Draveczky, 1966)	Middle

APPENDIX B

PERCENTAGE POINT DIFFERENCE BETWEEN EXPECTED AND OBSERVED PATTERNS OF SETTLEMENT DISTRIBUTION

<u>Variables</u>

Geology	A, Alluvium B, Loess C, Other
Soil	D, Skeletal, Lithomorph, and Less fertile forest soils E, Fertile forest soils F, Chernozem-brown forest intergrade G, Chernozem type soils H, Soils affected by water
Climate	I, Insufficient humidity over entire year or growing season K, Sufficient humidity
Vegetation	L, Deciduous forest M, Forest-steppe
Distance from rivers	O, Between 0 and 1 km P, Over 1 km

	Expected %	Körös %	Ppd	ALP %	Ppd	TLP %	Ppd
ADILO	0.42						
ADILP	0.15						
ADIMO	0.12					0.47	-0.35
ADIMP	0.09			0.26	-0.17		
ADKLO	0.64						
ADKLP	0.31						
ADKMO							
ADKMP							
AEILO	0.26						
AEILP	0.09						
AEIMO	0.09					1.89	-1.80
AEIMP	0.06						
AEKLO	0.23						
AEKLP	0.06						
AEKMO							
AEKMP							
AFILO	0.06						

(contd.)

	Expected %	Körös %	Ppd	ALP %	Ppd	TLP %	Ppd
AFILP	0.09						
AFIMO	0.09						
AFIMP	0.06						
AFKLO	0.20						
AFKLP	0.12						
AFKMO							
AFKMP							
AGILO	0.15						
AGILP	0.23						
AGIMO	1.45	6.10	-4.65	2.33	-0.88	0.47	
AGIMP	2.02	1.53		1.04			
AGKLO							
AGKLP							
AGKMO							
AGKMP							
AHILO	2.43			2.08		3.30	-0.87
AHILP	1.83						
AHIMO	7.83	22.14	-14.31	18.96	-11.13	4.25	
AHIMP	6.83	9.92	-3.09	5.71		0.47	
AHKLO	1.64			0.26		2.36	-0.72
AHKLP	0.47						
AHKMO							
AHKMP							
BDILO	0.17			0.52	-0.35		
BDILP	0.55			0.52		0.47	
BDIMO	0.09					1.42	-1.33
BDIMP	0.23			1.04	-0.81	0.47	-0.24
BDKLO	0.88			0.26		1.42	-0.54
BDKLP	2.35					2.36	-0.01
BDKMO	0.06						
BDKMP	0.04						
BEILO	0.20			0.52	-0.32	1.42	-1.22
BEILP	0.64			0.52			
BEIMO	0.45			0.52	-0.07	0.47	-0.02
BEIMP	1.07					2.83	-1.76
BEKLO	0.64					11.79	-11.15
BEKLP	1.34					5.19	-3.85
BEKMO	0.12						
BEKMP	0.23						
BFILO	0.28					0.94	-0.66
BFILP	0.36					0.47	-0.11
BFIMO	0.55			1.82	-1.27	0.47	
BFIMP	0.12			0.78	-0.66	0.47	-0.35
BFKLO	0.23					0.47	-0.24

(contd.)

	Expected %	Körös %	Ppd	ALP %	Ppd	TLP %	Ppd
BFKLP	0.53					0.94	-0.41
BFKMO							
BFKMP	0.06						
BGILO	0.23			1.04	-0.81	1.89	-1.66
BGILP	1.64			0.52		0.47	
BGIMO	3.33	17.56	-14.23	7.01	-3.68	1.89	
BGIMP	9.74	3.81		3.89		0.94	
BGKLO	0.12						
BGKLP	0.17						
BGKMO	0.17						
BGKMP	0.09						
BHILO	0.15			2.08	-1.93	0.47	-0.32
BHILP	0.12						
BHIMO	1.83	31.30	-29.47	20.00	-18.17	1.89	-0.06
BHIMP	1.89	2.29	-0.40	1.82			
BHKLO	0.17					1.89	-1.72
BHKLP	0.09						
BHKMO							
BHKMP							
CDILO	1.40			1.30		1.89	-0.49
CDILP	3.43			0.26			
CDIMO	0.83	0.76				2.83	-2.00
CDIMP	2.92					1.89	
CDKLO	3.65			3.63		10.85	-7.20
CDKLP	9.27			2.86		2.83	
CDKMO							
CDKMP	0.15						
CEILO	1.04			4.15	-3.11	1.42	-0.38
CEILP	2.26			2.86	-0.60		
CEIMO	0.61			1.30	-0.69	0.94	-0.33
CEIMP	1.34			0.52		3.30	-1.96
CEKLO	0.44					1.89	-1.45
CEKLP	1.26			0.26	-1.00	3.30	-2.04
CEKMO	0.06						
CEKMP	0.09						
CFILO	0.28			0.26		0.47	-0.19
CFILP	0.17						
CFIMO	0.28			2.08	-1.80	1.89	-1.61
CFIMP	0.80			0.26			
CFKLO	0.12					0.94	-0.82
CFKLP	0.15						
CFKMO							
CFKMP							
CGILO	0.09			1.05	-0.95	0.47	-0.38

(contd.)

	Expected%	Körös%	Ppd	ALP%	Ppd	TLP%	Ppd
CGILP	0.58						
CGIMO	1.21	1.53	-0.32	0.52		1.89	-0.68
CGIMP	3.22			0.26		0.94	
CGKLO							
CGKLP	0.03						
CGKMO							
CGKMP							
CHILO	0.83			2.34	-1.51	1.89	-1.06
CHILP	0.77			0.78	-0.01	0.47	
CHIMO	1.26	1.53	-0.27	1.30	-0.04	0.94	
CHIMP	1.75	1.53		0.52		0.47	
CHKLO	0.61					4.24	-3.63
CHKLP	0.28						
CHKMO							
CHKMP							
Total Ppd			66.74		48.96		53.61

ALP

	Early %	Ppd	Middle %	Ppd	Late %	Ppd
ADIMP					0.40	- 0.31
AGIMO			1.59	- 0.14	2.78	- 1.33
AGIMP			2.38	- 0.36	0.40	
AHILO			2.38		2.38	
AHIMO	20.68	-12.85	24.61	-16.78	17.06	- 9.23
AHIMP	17.24	-10.81	6.35		3.97	
AHKLO					0.40	
BDILO			0.79	- 0.62	0.79	- 0.62
BDILP					0.79	- 0.24
BDIMP					1.59	- 1.36
BDKLO					0.40	
BEILO					0.79	- 0.59
BEILP			0.79	- 0.15	0.40	
BEIMO			0.79	- 0.34	0.79	- 0.34
BFIMO			1.59	- 1.04	2.38	- 1.83
BFIMP					0.79	-0.67
BGILO			1.59	- 1.36	0.79	- 0.56
BGILP	3.45	- 1.81	0.79			
BGIMO			5.55	- 2.22	8.73	- 5.40
BGIMP			6.35		2.78	
BHILO	6.90	- 6.75	1.59	- 1.44	1.98	- 1.83
BHIMO	20.68	-18.85	24.61	-22.78	17.86	-16.03

(contd.)

	Early %	Ppd	Middle %	Ppd	Late %	Ppd
BHIMP	3.45	- 1.56	2.38	- 0.49	1.59	
CDILO					1.98	- 0.58
CDILP					0.40	
CDKLO			2.38		4.36	- 0.71
CDKLP					3.57	
CEILO	6.90	- 5.86	3.97	- 2.93	4.76	- 3.72
CEILP	6.90	- 4.64	1.59		2.78	- 0.52
CEIMO			1.59	- 0.98	0.79	- 0.18
CEIMP	3.45	- 2.11			0.40	
CEKLP					0.40	
CFILO					0.40	- 0.12
CFIMO	3.45	- 3.17	2.38	- 2.10	1.98	- 1.70
CFIMP					0.40	
CGILO	3.45	- 3.36			1.19	- 1.10
CGIMO			0.79		0.40	
CFIMP			0.79			
CHILO					3.57	- 2.74
CHILP	3.45	- 2.19	2.38	- 1.12	1.19	- 0.42
CHIMO	3.45	- 2.19	2.38	- 1.12	1.19	
CHIMP						
Total Ppd		73.96		54.85		52.13

TLP

	Middle %	Ppd	Late %	Ppd
ADIMO	0.63	- 0.51	1.59	- 1.47
AEIMO			6.35	- 6.26
AHILO	1.26		9.52	- 7.09
AHIMO	3.14		4.76	
AHKLO	3.14	- 1.50		
BDILP	0.63	- 0.08		
BDIMO			3.16	- 3.07
BDIMP			1.59	- 1.36
BDKLO	1.89	- 1.01		
BDKLP	3.14	- 0.79		
BEILO	1.89	- 1.69		
BEIMO			1.59	- 1.14
BEIMP	1.89	- 0.82	7.94	- 6.87
BEKLO	13.84	-13.20	3.16	- 2.52
BEKLP	6.92	- 5.58	1.59	- 0.25
BFILO	0.63	- 0.35	1.59	- 1.31

(contd.)

	Middle %	Ppd	Late %	Ppd
BFILP	0.63	- 0.27		
BFIMO			1.59	- 1.04
BFIMP			1.59	- 1.47
BFKLO	0.63	- 0.40		
BFKLP	1.26	- 0.73	1.59	- 1.06
BGILO	2.51	- 2.28		
BGILP	0.63			
BGIMO			6.35	- 3.02
BGIMP	0.63		1.59	
BHILO	0.63	- 0.48		
BHIMO	1.26		3.16	- 1.33
BHKLO	2.51	- 2.34	1.59	- 1.42
CDILO	1.89	- 0.49		
CDIMO	3.14	- 2.31	1.59	- 0.76
CDIMP	2.51		1.59	
CDKLO	12.58	- 8.93	6.35	
CDKLP	3.14		1.59	
CEILO	1.89	- 0.85	1.59	- 0.55
CEIMO	0.63	- 0.02	1.59	- 0.98
CEIMP	4.40	- 3.06	3.17	- 1.83
CEKLO	2.51	- 2.07		
CEKLP	3.77	- 2.51	6.35	- 5.09
CFILO			1.59	- 1.31
CFIMO	0.63	- 0.35	6.35	- 6.07
CFKLO	1.26	- 1.14	1.59	- 1.47
CGILO	0.63	- 0.54		
CGIMO	1.26	- 0.05	1.59	- 0.38
CGIMP	1.26			
CHILO	1.26	- 0.43	3.17	- 2.34
CHILP	0.63			
CHIMO	0.63		1.59	- 0.33
CHIMP	0.63			
CHKLO	5.66	- 5.05		
Total Ppd		59.83		61.79

LATE ALP

	Late AVK %	Ppd	Tiszadob %	Ppd	Esztar %	Ppd
ADIMP	1.23	- 1.14	1.49	- 1.40		
AGIMO			1.49	- 0.04	5.55	- 4.10
AGIMP					2.78	- 0.76
AHILO	1.23				8.33	- 5.90
AHIMO	18.53	-10.70	8.96	- 1.13	25.00	-17.17
AHIMP	2.47		1.49		11.11	- 4.28
BDILO	1.23	- 1.06	1.49	- 1.32		
BDILP	1.23	- 0.68	1.49	- 0.94		
BDIMP	1.23	- 1.00	5.97	- 5.74		
BDKLO	1.23	- 0.35	1.49	- 0.61		
BEILO			2.99	- 2.79		
BEILP	1.23	- 0.59	1.49	- 0.85		
BEIMO	2.47	- 2.02	2.99	- 2.54		
BFIMO	4.95	- 4.40	2.99	- 2.44		
BFIMP	1.23	- 1.11	2.99	- 2.87		
BGILO					5.55	- 5.32
BGIMO	6.18	- 2.85	1.49		5.55	- 2.22
BGIMP	2.47		1.49			
BHILO	1.23	- 1.08	4.48	- 4.33		
BHIMO	17.29	-15.46	2.99	- 1.16	2.78	- 0.95
BHIMP	1.23		1.49		5.55	- 3.66
CDILO	3.71	- 2.31	4.48	- 3.08		
CDKLO	8.65	- 5.00	7.46	- 3.81		
CDKLP	7.41		7.46			
CEILO	2.47	- 1.43	7.46	- 6.42	5.56	- 4.52
CEILP	1.23		4.48	- 2.22	5.56	- 3.30
CEIMO	1.23	- 0.62	1.49	- 0.88		
CEIMP					2.78	- 1.44
CEKLP			1.49	- 0.23		
CFILO	1.23	- 0.95	1.49	- 1.21		
CFIMO	3.71	- 3.43	4.48	- 4.20		
CFIMP			1.49			
CGILO					5.56	- 5.47
CGIMO					2.78	- 1.57
CHILO	2.47	- 1.64	5.98	- 5.15		
CHILP					5.56	- 4.79
CHIMO	1.23		1.49	- 0.23		
CHIMP			1.49			
Total Ppd		57.82		55.59		65.45

Late ALP

	Szilmeg		Bükk		Szakálhát	
	%	Ppd	%	Ppd	%	Ppd
ADIMP			1.28	- 1.19		
AGIMP	3.57	- 1.55			4.88	- 2.86
AHILO			2.56	- 0.13		
AHIMO	28.57	-20.74	8.97	- 1.14	23.17	-15.34
AHIMP	7.15	- 0.32	2.57		6.10	
AHKLO			1.28			
BDILO			1.28	- 1.11		
BDIMP			2.57	- 2.34		
BDKLO			1.28	- 0.40		
BEILO			1.28	- 1.08		
BEILP			1.28	- 0.64		
BFIMO			3.85	- 3.30		
BFIMP			2.57	- 2.45		
BGILO	7.15	- 6.92				
BGIMO					19.51	-16.18
BGIMP	3.57		1.28		3.66	
BHILO			6.41	- 6.26		
BHIMO	10.72	- 8.89	5.13	- 3.30	39.02	-37.19
BHIMP	7.14	- 5.25	2.57	- 0.68	1.22	
CDILO			5.13	- 3.73		
CDILP			1.28			
CDKLO			11.54	- 7.89		
CDKLP	3.57		7.69			
CEILO	3.57	- 2.53	8.97	- 7.93		
CEILP	3.57	- 1.31	6.41	- 4.15		
CEIMO	3.57	- 2.96			1.22	- 0.61
CEIMP	3.57	- 2.23	1.28			
CFIMO			2.57	- 2.29		
CGILO	7.14	- 7.05				
CHILO			7.69	- 6.86		
CHILP	7.14	- 6.37				
CHIMO			1.28	-0.02		
CHIMP					1.22	
Total Ppd		66.12		56.89		72.18

www.ingramcontent.com/pod-product-compliance
Lightning Source LLC
LaVergne TN
LVHW072252300926
849160LV00018B/1647